ARISTOTLE'S *PHYSICS* BOOK I

This book provides a comprehensive and in-depth study of *Physics* I, the first book of Aristotle's foundational treatise on natural philosophy. While the text has inspired a rich scholarly literature, this is the first volume devoted solely to it to have been published for many years, and it includes a new translation of the Greek text. Book I introduces Aristotle's approach to topics such as matter and form, and discusses the fundamental problems of the study of natural science, examining the theories of previous thinkers including Parmenides. Leading experts provide fresh interpretations of key passages and raise new problems. The volume will appeal to scholars and students of ancient philosophy as well as to specialists working in the fields of philosophy and the history of science.

DIANA QUARANTOTTO is Senior Research Fellow in the Department of Philosophy at Sapienza University of Rome. Her work focuses on Aristotle's physics, metaphysics and biology, and she is the author of *Causa finale, sostanza, essenza in Aristotele* (2005) and of *L'universo senza spazio. Aristotele e la teoria del luogo* (2017).

ARISTOTLE'S *PHYSICS* BOOK I

A Systematic Exploration

EDITED BY

DIANA QUARANTOTTO

Sapienza University of Rome, Italy

CAMBRIDGE
UNIVERSITY PRESS

CAMBRIDGE
UNIVERSITY PRESS

University Printing House, Cambridge CB2 8BS, United Kingdom

One Liberty Plaza, 20th Floor, New York, NY 10006, USA

477 Williamstown Road, Port Melbourne, VIC 3207, Australia

314–321, 3rd Floor, Plot 3, Splendor Forum, Jasola District Centre, New Delhi – 110025, India

79 Anson Road, #06–04/06, Singapore 079906

Cambridge University Press is part of the University of Cambridge.

It furthers the University's mission by disseminating knowledge in the pursuit of education, learning, and research at the highest international levels of excellence.

www.cambridge.org
Information on this title: www.cambridge.org/9781107197787
DOI: 10.1017/9781108181853

First published 2018

Printed in the United Kingdom by Clays, St Ives plc

A catalogue record for this publication is available from the British Library.

ISBN 978-1-107-19778-7 Hardback

Contents

Contributors

ISTVÁN BODNÁR is Professor of Philosophy at Eötvös University, Budapest, and is Faculty Member at the Central European University, Budapest.

LAURA M. CASTELLI is Research Fellow (DFG) at the Munich School of Ancient Philosophy, LMU Munich.

CRISTINA CERAMI is a Senior Research Fellow (CR1) of the French Centre Nationale de la Recherche Scientifique (CNRS), UMR 7219.

DAVID CHARLES is Howard H. Newman Professor of Philosophy at Yale University.

TIMOTHY CLARKE is Assistant Professor of Philosophy at the University of California, Berkeley.

ALAN CODE is Ward W. and Priscilla B. Woods Professor of Philosophy and Professor of Classics at Stanford University.

ANDREA FALCON is Associate Professor at Concordia University, Montreal.

LINDSAY JUDSON is Associate Professor of Philosophy at Christ Church, University of Oxford.

JAMES G. LENNOX is Professor of History and Philosophy of Science at the University of Pittsburgh.

DIANA QUARANTOTTO is Senior Research Fellow at Sapienza University of Rome.

Preface

Physics I is the first book of the *Physics*, Aristotle's basic treatise on natural philosophy. It is the beginning of his scientific inquiry into natural things: it outlines the first principles of such things (i.e. matter and form) and discusses preliminary and fundamental problems for any such an inquiry (for instance, is natural science possible at all?). The book is designed to connect Aristotle's own project with the previous tradition with the aim of overcoming its main difficulties and resolving the most serious mistakes of earlier philosophers. To this end, his investigation into principles is intertwined with a doxographical inquiry into the theories of some of his predecessors. The outline that Aristotle sketches in *Physics* I is developed in the subsequent books of the *Physics* and in other treatises on natural science and metaphysics.

Given its central role in Aristotle's science of nature and his philosophy as a whole, *Physics* I has inspired a rich scholarly literature. However, no recent monograph or collection of papers has been devoted solely to it. With the aim of filling this gap, I organised an international conference on this book, inviting some of the leading specialists in these aspects of Aristotle's philosophy. The conference took place in September 2013 at Sapienza University of Rome, and was followed by a second one there in June 2015. I heard, while in the process of organising the first, that the committee of the Symposium Aristotelicum was planning to hold its next session in Greece on the same subject. This coincidence seemed opportune since, given the especially large group of scholars thinking about *Physics* I, there was bound to be much interest in the book as well as ideas for more than one volume. It also seemed appropriate that the two conferences be held in Greece and Rome, the two centres of the ancient world.

The preparatory work for the Roman conferences and the discussions that arose from them have resulted in this volume, which provides a comprehensive and in-depth study of *Physics* I. The volume consists in a general introduction, nine essays, each giving a close textual reading of one of *Physics* I's nine chapters, and a fresh translation of the Greek text, where each contributor offers a new translation of his or her own chapter, which Lindsay Judson and I have coordinated. Our project has been motivated by four main exegetical goals: to uncover Aristotle's argument as it unfolds through the book; to understand his main theoretical commitments; to reconstruct his argumentative strategies, assumptions and background; and to assess the epistemic status of the inquiry conducted in *Physics* I and its relation with subsequent investigations into physics. Fresh interpretations of several key passages are proposed and several new problems raised.

My introductory essay addresses the general issues of the role, structure and status of *Physics* I. First, I seek to clarify the sense in which *Physics* I is the beginning of Aristotle's physical project. I argue that, although *Physics* I's inquiry is of a scientific and physical kind, it is a relatively free-standing treatise, which plays an introductory role aimed at, as it were, setting the scene for the project as a whole. I highlight various clues that show *Physics* I's introductory role: the way in which it describes the central object of natural science (i.e. natural substances), the evidence it uses, its heuristics, its peculiar relation with the tradition of natural philosophy, the way it identifies and describes the principles of natural substances. In order to clarify further the role of *Physics* I, I explore the relation between this book and other parts of the corpus. In particular I consider *Physics* I's assumptions and background (by focusing especially on *Physics* I's relations with the *Organon*) and the connections between *Physics* I and what comes next in Aristotle's physical enterprise. I also focus on the overall argumentative strategy of the book, addressing its various schemes of construction and the order in which the principles are introduced in *Physics* I–II. I argue that the particular mode of inquiry adopted in *Physics* I–II is especially useful, perhaps required, in constructing a correct theory of principles. Further, I suggest that the principles argued for by Aristotle in *Physics* I are the essences of the things that play the role of principles as such: they are the *ti esti* of principles, and this is why they are *first* principles.

In the first of the following essays, Andrea Falcon examines the first chapter of *Physics* I, where Aristotle outlines his method for discovering the principles of natural things. This method is described as a natural road that starts from what is confounded to a degree (i.e. humans' initial grasp of natural things) and leads to its division into parts, i.e. into its first principles (which the subsequent inquiry will show to be matter and form). In his essay, Falcon focuses on three main questions: how are principles of natural science introduced; what is the precise nature of the road to the principles; how do the two examples at the end of the chapter contribute to illustrating the search for principles? He notes that, in *Physics* I.1, Aristotle assumes the existence of principles and conceives of them broadly as principles of knowledge. In this connection he emphasises, more generally, that, in outlining the task that lies ahead in *Physics* I, Aristotle is casting his net as wide as possible. Concerning the road to the principles, Falcon argues that it is a natural and necessary path, which includes three steps, each one building on the previous one, and that the epistemic journey towards the principles is a rational process. Lastly, he suggests that the first of the two examples given at the end of the chapter is intended to illustrate the existence of two different epistemic stages, while the second shows the progression from one to the other.

After chapter 1's methodological introduction, the second chapter, discussed by Timothy Clarke, starts by classifying the theories of principles of some of Aristotle's predecessors. As Clarke suggests, this classification provides one of the main frameworks for the book's subsequent investigation. In this way, the epistemic journey towards the principles, outlined in chapter 1, is intertwined with a doxographical inquiry. This inquiry – even though not conducted in a chronological order – seems aimed at retracing and assessing the steps of Aristotle's predecessors along the road towards the principles, and especially at showing and rectifying their mistakes. One of the main focuses of this inquiry is the distinction between a property and the bearer of a property, which is among the principal means that, in *Physics* I, Aristotle uses to draw his distinction between matter, conceived of as that which underlies, and form.

In his doxographical inquiry, Aristotle first considers the Eleatic claim that 'what is is one and changeless'. The Eleatic theory implicitly denies that it is possible to apply the method outlined in chapter 1: the division of what is confounded into parts. In Aristotle's view, this theory is neither about nature (since it denies change and plurality) nor about principles (since it denies plurality, and hence the distinction between a principle and the thing it is a principle of). As a result, its examination lies outside the scope of natural science. Nonetheless, he devotes a considerable amount of time to the Eleatics in *Physics* I, since they raise difficulties relevant to natural science. He focuses on Eleatic monism in chapters 2 and 3, and on Eleatic immobilism in chapter 8.

In chapter 2 Aristotle provides his refutation of Eleatic monism. In his essay, Clarke argues that Aristotle interprets the Eleatics as advancing two radical kinds of monism: entity monism (the view that reality consists of just one entity) and essence monism (the view that reality is all of the same essence or nature). Clarke examines Aristotle's attempt to show that neither of these positions can coherently be maintained.

After arguing, in chapter 2, that Eleatic monism is impossible, Aristotle goes on to explain in chapter 3 why the Eleatics' arguments are unsuccessful. This chapter is discussed by Laura M. Castelli. In her essay, she analyses Aristotle's grounds for disputing the premises and the logical structure of Melissus' and Parmenides' arguments, and gives an account of the way in which the analysis of such arguments is relevant to his inquiry into nature. In particular, she emphasises that Aristotle's initial criticism of Parmenides at 186a22–32 is based on a certain understanding of 'signification', and that 186a34–b14 presents arguments that a monist might be willing to accept in order to avoid the initial criticism based on Aristotle's distinction between a property and its bearer.

Aristotle, after concluding his criticism of Eleatic monism with the claim that 'it is impossible that what is is one in this way', turns in chapter 4 to the theories of principles of the physicists. Unlike the Eleatic theory, these are true theories of principles (since they admit plurality and conceive of some things as principles of other things), and are about nature (since they admit change and try to explain it). Further, they follow, at least up to a point, the method outlined in chapter 1: the division of what is confounded (i.e. a natural substance) into parts. What they lack is a correct application of this method. Hence they fail to divide correctly natural substances into their true first principles. Aristotle employs his doxographical investigation to move towards his own theory of principles, and especially its cornerstone: the notion of that which underlies (i.e. matter).

Chapter 4, examined by Cristina Cerami, starts with a new formulation of the classification of theories of principles given in chapter 2, and provides a criticism of Anaxagoras' view that principles are infinite and that generation occurs by aggregation and separation. In her essay, Cerami argues that Aristotle's analysis and refutation of Anaxagoras' theory are of major importance for Aristotle's project in *Physics* I. According to Cerami, in Aristotle's view Anaxagoras acknowledged a certain kind of distinction between that which underlies and its determinations. However, his positive account exemplifies the inconsistency in those theories that do not allow for a proper separation between predicates and that which underlies. By attacking Anaxagoras' theory of an infinite number of principles, Aristotle aims to show that separation is necessary for the explanation of change, but that this notion must be qualified. If we admit an infinite number of principles, it turns out that, contrary to what Anaxagoras himself believed, we cannot account for the existence of a plurality of knowable things. Separation cannot proceed *ad infinitum*. Indeed, with a proper understanding of this notion, we can see that separation must stop at some point, due to the fundamental distinction between a property and its bearer.

From chapter 5 onwards Aristotle's inquiry takes a different course. It does not proceed by criticising historically held views, but initiates a positive investigation of the number and nature of the principles. Chapters 5 and 6 provide the positive basis for the theory of principles that Aristotle formulates in chapter 7. They arrive at the preliminary conclusion that principles are either two (two opposites) or three (two opposites and one underlying thing), and present the choice between these alternatives as a puzzle, which will be solved in chapter 7.

In chapter 5, analysed by Lindsay Judson, Aristotle starts from the general opinion (*endoxon*) that opposites are principles – which was already introduced in chapter 4 – and provides arguments to support it. In returning to the subject of the theory of principles of his predecessors, he raises the issue of what the first principles (in this case, the first opposites) are – an issue that was possibly mentioned first in chapter 1 (184a13–14). In his essay, Judson discusses the main issues of chapter 5: how we should understand, and best translate, the term ἐναντίον; how we should understand Aristotle's various claims about principles (ἀρχαί); the tension between the claims that ἐναντία are not from each other but come to be from each other; questions about the logical grammar and/or the ontological commitments of these claims, about what the principles in question are principles of, and about Aristotle's insistence that these principles are *opposites*. In conclusion, Judson provides an assessment of Aristotle's position at the end of chapter 5.

In chapter 6, which is discussed by Alan Code, Aristotle assumes that the principles are opposites, introduces the underlier as a further principle, and investigates the number of the first principles. He narrows down the number of the first principles to just two or three, and concludes by saying that the question as to whether they are two or three is very difficult. Code provides an analysis of the four phases of this inquiry: arguments against there being just one principle or an unlimited number; arguments for the underlier as a third principle that is not an opposite; discussion of previous theories that had such a third principle; and arguments against there being more than one principle that is an underlier, or more than two principles that are opposites. The narrowing down of the number of opposites to two, the introduction of a third principle that is not an opposite, and the claim that there is just one principle that plays the role of underlier are achieved by developing the inquiry at a level of maximal generality (i.e. first the level of kinds, and then the level of substance, conceived of as a kind of being), and by employing the distinction between the being of a property and the being of the bearer of a property. Code emphasises that the results of chapter 6 are only an intermediate step towards Aristotle's theory of principles. Once the third principle has been introduced, it needs to be incorporated into the account of the principles that has progressively emerged, and this incorporation will need modifications, refinements and adjustments. The concluding difficulty (*aporia*) about whether the principles are two or three both reflects how hard this is and is a challenge whose resolution leads to his own positive account of the principles.

This positive account of the principles and the solution of the *aporia* raised at the end of chapter 6 (i.e. whether the principles are two or three) are finally provided in chapter 7, which is discussed by David Charles. In chapter 7, Aristotle claims that the principles of natural things are two in a sense (that which underlies and form: 190b17–20, or the two opposites: 190b29–32), and three in another (that which underlies, privation and form: 190b35–191a3). Charles argues that, in this chapter, Aristotle makes three important moves. He introduces and partially elucidates the idea of that which underlies, understood as something that remains through the coming to be of certain objects. Second, he argues that there is something which underlies in this way in all cases of the coming to be. Third, he suggests that what underlies is not only a principle of the coming to be of objects but also of their continued existence: there is one continuing thing which remains from the beginning of the process of generation of an object to its final destruction. In assessing Aristotle's position in chapter 7, Charles stresses that Aristotle's introduction of that which underlies is intended to pick out something which will subsequently be analysed more fully in terms of matter and potentiality. Charles also considers some specific issues about Aristotle's arguments in chapter 7, which raise more general questions about his methodology.

Aristotle, after setting out his own account of the first principles of natural substances in chapter 7, returns to the Eleatic theory, in chapter 8, and in particular to the Eleatic denial of change. He makes the momentous claim that an Eleatic puzzle against generation can be resolved only through the principles he has established. In his essay on chapter 8, István Bodnár argues that the discussion of this Eleatic *aporia* clarifies the status and interdependence of the underlier and privation, with the conclusion that the explanatory framework of earlier philosophers fails to do justice to this interplay. It turns out that, in Aristotle's view, only Plato touched on the nature of the underlier – though not even he grasped it adequately – and sought to provide a possible alternative response to the Parmenidean puzzle about change.

In the ninth and concluding chapter, examined by James G. Lennox, Aristotle criticises the Platonic alternative, which identifies matter with opposition and form with unity – a view already alluded to in chapters 4 and 6. Lennox argues that Aristotle was familiar with this account from his time in the Academy and at Assos. He sees chapter 9 as Aristotle's presentation and rejection of this alternative through an extended comparison of his own view of the relation between matter, form and privation as principles of natural change and the Academic alternative – a presentation often self-consciously worded in ways that will make the issues clear to his Academic critics. Further, Lennox provides an interpretation of this last chapter of *Physics* I as designed as a transition to book II. In Lennox's view, chapter 9 prepares the way for further discussion on whether matter or form is the most suitable candidate for the nature of natural things, and thus for a radically different investigation of the primary causes and primary principles of natural things. This transition is emphasised by the concluding words of the book: 'So, then, let us take it as determined in this manner that there are principles and which and how many in number they are; but let us carry on the discussion, beginning from another starting point.'

The first book and first beginning of Aristotle's *Physics* ends here. This book not only sets the scene for Aristotle's physical project, but also – even if built as a relatively free-standing treatise – gives some hints about how that project will develop.

Acknowledgements

I would like to express my gratitude to all the contributors for joining the project and for their active participation and exciting interplay. My special thanks go to the following: Andrea Falcon for his generous help in the preparation of the book project; Lindsay Judson for coordinating the translation with me and for liberally sharing his ideas; David Charles for his advice, support and exchange of ideas. In addition, I would like to thank Anna Maria Ioppolo for her long-standing support, Maria Cristina Dalfino for her invaluable help in the final editing and indexing, and Sapienza University of Rome for funding the conference that started this ball rolling. Finally I am grateful to my family for the patience and support they have shown me during this project.

Abbreviations

Alex. *In Metaph.*	Alexander Aphrodisiensis, *In Aristotelis metaphysica commentaria*
APo	Aristoteles, *Analytica posteriora*
APr	Aristoteles, *Analytica priora*
Averr. *LC Phys.*	Averrois Cordubensis, *Commentarium Magnum In Aristotelis De Physico Audito*
Cat.	Aristoteles, *Categoriae*
DA	Aristoteles, *De anima*
DC	Aristoteles, *De caelo*
DK	H. Diels and W. Kranz, *Die Fragmente der Vorsokratiker*
GA	Aristoteles, *De generatione animalium*
GC	Aristoteles, *De generatione et corruptione*
HA	Aristoteles, *Historia animalium*
Int.	Aristoteles, *De interpretatione*
Metaph.	Aristoteles, *Metaphysica*
Meteor.	Aristoteles, *Meteorologica*
MXG	Aristoteles, *De Melisso, Xenophane et Gorgia*
NE	Aristoteles, *Ethica Nicomachea*
OCT	*Oxford Classical Texts*
PA	Aristoteles, *De partibus animalium*
Parm.	Plato, *Parmenides*
Phil.	Plato, *Philebus*
Philop. *In Phys.*	Iohannes Philoponus, *In Aristotelis physicorum libros commentaria*
Phys.	Aristoteles, *Physica*
SE	Aristoteles, *Sophistici elenchi*
Simpl. *In Phys.*	Simplicius, *In Aristotelis physicorum libros commentaria*

Soph.	Plato, *Sophista*
Themist. *In Phys.*	Themistius, *In Aristotelis physica paraphrasis*
Tim.	Plato, *Timaeus*
Top.	Aristoteles, *Topica*

The Role, Structure and Status of Aristotle's *Physics* I

Diana Quarantotto

I. *Physics* I: A Beginning

Physics I is the beginning of Aristotle's scientific inquiry into natural things. Its declared aim is to 'try to determine the things concerning the principles' (184a14–16). Aristotle stresses in various ways that the book is a beginning. In chapter 1 he describes its aim as the first thing that the natural scientist must try to accomplish (ibid.). Moreover, he presents the inquiry that he is going to conduct as the start of an epistemic journey (184a16–26). Further, at the end of the book he describes the investigation he has just carried out as a beginning to be followed by another beginning (192b4): *Physics* I is the first of two beginnings.

But 'beginning' is said in several senses. So, to understand what *Physics* I is about, we should clarify what kind of beginning it is and what its role is within Aristotle's physical project. I shall start by considering *Physics* I's epistemic status and specific approach to the issues it deals with.

First of all, *Physics* I is a beginning that displays clear signs of continuity both with what follows it in the *Physics* and, more generally, with the corpus of Aristotle's physical writings. It is to a large extent a scientific physical investigation. Indeed, as already mentioned, its aim is described as the first thing that the natural scientist must try to accomplish. Of course, it is not a piece of *episteme* in the sense of *APo* (i.e. demonstrative science), since it is not aimed at constructing demonstrations from principles, but rather at finding the principles needed to construct demonstrations. However, it is clearly presented as an instance of physical science. The only exception is the inquiry into Eleatic monism and immobilism that

I would like to thank the participants in the meetings on *Physics* I held in Roma, Sapienza (September 2013, June 2015) for their helpful comments. I am especially indebted to David Charles, Timothy Clarke, Alan Code, Andrea Falcon and Lindsay Judson. I am beholden to the participants of the Seminario itinerante di filosofia antica held in various Italian cities in 2013 and also to the participants of the Symposium Aristotelicum held in Delphi in 2014 for further discussion of these issues.

occupies part of chapter 2, chapter 3 and chapter 8. This inquiry is described as something that lies outside the natural scientist's expertise (184b25–185a20). But this exclusion, by contrast, implies that the rest of the investigation carried out in the book does belong to the natural scientist's expertise. The continuity of *Physics* I with the rest of the *Physics* and especially with books II–IV is also shown by the content and method of these investigations. *Physics* I starts an inquiry into the principles of natural things, which is carried forward in the subsequent books: it introduces and focuses on two of the four causes of natural substances (matter and form),[1] whereas book II, besides inquiring further into matter and form, and besides defining 'nature', tackles the other two causes (mover and goal),[2] and the subsequent books deal with issues whose treatment is required to understand more clearly what nature is (e.g. change and the infinite in book III; place, void and time in book IV). From the methodological point of view these inquiries do not differ significantly: they all employ a combined method, which comprises an inductive procedure and a doxographical inquiry, and use evidence and *phainomena* of various kinds.[3]

The continuity of *Physics* I with the other physical investigations is also suggested by the way in which Aristotle refers to *Physics* I elsewhere, for instance in *Meteor.* I.1 and in several passages of the *Metaphysics.* The first entry of the *curriculum* of the physical investigation presented in *Meteor.* I.1 seems to include the inquiry into principles conducted in *Physics* I: 'We have already discussed the first causes of nature, and all natural motion' (338a20–1). Further, some passages from the *Metaphysics* are even more telling because their reference to *Physics* I is more precise. Here Aristotle refers to *Physics* I without showing any sign of discontinuity between it and the other physical *logoi.* This suggests that he considers it either an instance of the *logoi peri physeos* or an integral part of them (*Metaph.* A 3.983a33–b1, A 5.986b30–1, A 7.988a21–2, A 10.993a11–13, K 6.1062b31–3, M 9.1086a21–4).[4]

1 *Phys.* I only hints at the role of form as mover and as goal (191a3–7, 192a16–25). Nevertheless, it thereby prepares the ground for *Phys.* II's inquiry into these principles (on this connection, see Lennox in this volume).

2 In section II, I come back to the issue about the relation between *Phys.* I and II, and the order in which the principles are introduced in these two books.

3 It should be added, however, that *Phys.* I stands out for a particular use of dialectical principles and distinctions. I come back to this issue in section III.

4 *Metaph.* M 1.1076a8–9 may be added to this list, with the caveat that it might refer to *Phys.* II.

These are some of the most obvious signs of continuity between *Physics* I and the other physical investigations. But *Physics* I also presents some other features that, without denying its general physical character, offer a different perspective as well as further clues to its specific role. The features I shall focus on make *Physics* I look like a relatively free-standing treatise, aimed at introducing the science of nature by, as it were, setting its scene, by working out, at some preliminary level, the principles of natural substances. I shall progressively clarify the sense in which *Physics* I plays an introductory role.

Physics I's inquiry begins and carries on to its conclusion without providing a definition of the subject matter of natural science (i.e. nature and natural things). Aristotle gives a first description of this subject matter from the point of view of how it appears at the beginning of the inquiry ('what is more knowable to us'): a 'confounded whole', a 'universal' and something perceptible (184a16–26). Then he describes it generically as a multitude of entities, some or all of which undergo change (185a12–14). These descriptions do not refer exclusively to natural entities. Moreover, the inquiry uses evidence that either does not concern natural entities alone or concerns aspects of natural entities that are not peculiar to them. Therefore, the principles arrived at as a result of the investigation, although presented as principles of natural substances, are also applicable to other kinds of things: they are applicable to everything that changes, comes to be and passes away, regardless of whether these processes are natural or not, and whether what undertakes them is a natural thing or not. Indeed, the distinction between natural and non-natural things is not drawn until *Phys.* II.1, where Aristotle defines nature as an internal principle of change and rest. Moreover, in *Phys.* II.2 a further refinement of the concepts of form and matter will also be used to distinguish the specific subject matter of physics from those of mathematics, first philosophy and the scientific enterprises set up by Aristotle's predecessors. From this point of view, the principles tackled in *Physics* I (form and matter), in so far as they do not hold exclusively of natural entities and are not sufficient to distinguish the subject matter of physics from that of other sciences or from things that are not among the objects of natural science (i.e. artefacts), are not yet distinctively physical principles.[5] This is one of the senses in which *Physics*

[5] This does not mean that *Phys.* I does not have a distinctively physical goal: indeed, the book is explicitly aimed at finding the principles of natural science. By 'non distinctively physical principles' I mean principles that, by themselves, are not sufficient to distinguish natural from non-natural substances, and that are introduced by means of evidence that is not only physical.

I may be considered introductory: it prepares the ground for the investigations carried out from *Physics* II on, and it does so by outlining and focusing on the notions of matter and form, which, together with those of mover, goal and means, will be used in *Physics* II both to define the central object of natural science (II.1), and to delimit the domain of physics (II.2). The kind of introduction provided in *Physics* I may have heuristic and/or didactic purposes. Indeed, nature is difficult and the principles of natural things are especially tricky to disentangle. So, Aristotle may have thought it necessary, or at least highly advisable, to start the search for principles by inquiring about a broad sample of things and at a level in which the distinction between form and matter is easier to draw. Having done that, in *Physics* II he goes on to show that, in the case of natural substances, form and matter play also the role of internal principles of change, and that they stand in a teleological relation in which form is the goal and matter is the means to it.

Besides starting from a description of its subject matter that is not exclusive to natural things and introducing principles that are not distinctively physical (in the sense just mentioned), *Physics* I starts from a description of its subject matter (i.e. of how the subject matter appears at the beginning of the inquiry) that is generic from the ontological point of view. It starts from the very broad idea that this subject is a perceptible whole comprising many things as its parts (184a24–6) or some sort of combination of things that are confused together (184a21–2). In this respect, the starting point of the investigation initiated in *Physics* I differs significantly from that of, for instance, *DC* I.1 or III.1.[6] Here, Aristotle introduces the inquiry by providing a description of the ontological status of its central object that is much more precise and employs a distinction between substance and affections that belong to the substance: physics, claims Aristotle, deals with bodies, or sensible substances, magnitudes, and with their affections and changes (268a1–6, 298a27–b5). What in *De caelo* is a starting point of the inquiry, in *Physics* I is instead part of its final achievement.[7] One of the main aims of *Physics* I is to characterise the central object of natural science from the ontological point of view as a substance, conceived of as a compound of form and matter (190b17–23) receptive of accidental affections, and to present this characterisation as an

[6] On a connected point concerning Theophrastus' approach to this issue, see Falcon in this volume.

[7] This of course is not to say that the description of the central objects of physics as bodies or bodily substances affected by *pathe*, provided in *DC* I.1 and III.1, is exactly the same as *Phys.* I's description of them as compounds of form and matter.

alternative to those provided by Aristotle's predecessors (both the Eleatics, who conceive of what is as an absolute and indivisible unity, and maintain that natural things are only appearances or false appearances,[8] and the physicists, who conceive of the ontological composition of natural substances in ways that are different from Aristotle's). Further, in *Physics* I, Aristotle's project of characterising the ontological structure of the central object of physics amounts to determining its principles and, first of all, to determining that there are principles of natural things, i.e. that natural things can be divided into principles (184a22–3).[9] *Physics* I plays an introductory role not only because it introduces principles that are not distinctively physical, but also because it outlines the ontological status of the object of natural science in terms that are not distinctively physical.

Another sign of the introductory character of *Physics* I is found in the fact that the book tackles at length two basic starting points of natural science: the existence of both change and plurality (185a12–14). The treatment of this issue is a non-physical or a pre-physical inquiry since, in Aristotle's view, the natural scientist derives these starting points from sense-perception and takes them as obvious truths, without having to account for them or to argue against those who deny them (185a14–17). Aristotle tackles this issue by addressing and criticising the Eleatics' denial of change and plurality, a denial that caused them to deviate from what in Aristotle's view is the natural path of the inquiry into principles (191b31–3) and thus bringing them outside the domain of natural science (184b25–185a1) and, even more radically, of any science (185a1–5).[10]

8 *Phys.* VIII.3.254a27.

9 For a description of the aim of *Phys.* I (and of some of the inquiries of Aristotle's predecessors) in ontological terms, which is presented as equivalent to the aim of searching for the principles of natural things, cf. 184b22–5 and below, section II.

10 In *Phys.* I Aristotle considers the Eleatic theory from two different points of view: the point of view of the epistemological structure of natural science and that of its historical development. From the first point of view, the Eleatic theory is not about nature (184b25–185a1, 185a17–18), since it denies the basic starting points of natural science (i.e. the existence of change and plurality). Moreover, the Eleatic claim that what is is absolutely one denies one of the fundamental presuppositions of any science: the existence of principles, conceived of as distinct (in some way) from what they are principles of (185a1–5). However, when natural science is considered from the perspective of its historical development, Aristotle describes the Eleatics as philosophers who inquired into nature (191a24–5) and who deviated from the natural path of this inquiry (191a26–7, 191b31–3) because of physical difficulties (185a18) they were unable to solve. In other words, the Eleatics' inquiry is seen as an argument, or a series of connected arguments, that is about nature or that raises physical puzzles, but whose final conclusion that what is is one and changeless is not about nature. This latter view on the Eleatics is employed also in *Metaph.* A 3.984a16–b1.

Physics I is not the only text in the *Physics* where Aristotle addresses the pre-physical issues connected with the starting points of natural science and with the Eleatic denial of nature. Other relevant texts are, for instance, *Phys.* II.1 (193a3–9) and especially *Phys.* VIII.3. However, *Physics* I stands out for the breadth and complexity of its investigation of these issues.[11] Some aspects of the particular strategy used in *Physics* I to tackle the starting points of physics and the Eleatic denial of nature repay closer examination in order to get a clearer idea of *Physics* I's specific role, i.e. of the kind of relevance for natural science that Aristotle discerns in his treatment of these pre-physical issues in *Physics* I.

Aristotle's treatment of the Eleatic theory in *Physics* I is aimed specifically at resolving the Eleatic arguments, rather than refuting their claims (185a8–10, 186a5–11, 22ff.).[12] In doing this, Aristotle is not demonstrating the starting points of physics, i.e. that nature or multiplicity or change exist, but detecting and removing the mistakes that brought the Eleatics and their heirs to deny these obvious facts. This is relevant for at least three reasons.

First, Aristotle maintains that it would be ridiculous to try to demonstrate obvious facts, attested by sense-perception, like the existence of nature, of plurality and of change. For it would imply proving what is obvious by what is obscure (193a3–9). Hence, the only acceptable or

[11] *Phys.* II.1's and VIII.3's treatments of this issue are much shorter and simpler than *Phys.* I's. *Phys.* II.1 labels as ridiculous the attempt to prove that nature exists and provides reasons for this judgement. *Phys.* VIII.3 tackles only the Eleatic denial of change, and describes this task with words that are very similar to those used in *Phys.* I.2, but its overall strategy is very different from the one adopted in *Phys.* I. The main difference is that *Phys.* VIII.3 provides a refutation of the Eleatic thesis and not, like *Phys.* I, also a resolution of the Eleatic arguments. It is noteworthy that neither in *Phys.* II.1 nor in *Phys.* VIII.3 does Aristotle make any reference to *Phys.* I. One may try to speculate over the reasons why in *Phys.* II.1 and VIII.3 (unlike for instance in *Metaph.* A 5.986b10–31, which is an interesting parallel passage) Aristotle does not mention *Phys.* I. The connections between *Phys.* I and both of these passages is sufficiently strong to justify such a speculation. One possible reason is that *Phys.* II.1 and/or *Phys.* VIII.3 were originally written before *Phys.* I and so ignore the inquiry conducted there. A clue in this direction may be provided by *Phys.* II.1.193b20–1. There Aristotle says that it still has to be determined whether there is privation and the opposite within absolute generation, which is an issue that he clearly addresses in *Phys.* I and, at least to a certain extent, considers there as established. In so far as *Phys.* II.1 is concerned, another possibility, not necessarily incompatible with the first, is that, unlike some parts of *Phys.* I, *Phys.* II.1 is considered by Aristotle as a distinctively physical investigation, which therefore does not deal with the starting points of natural science, since it would amount to transcending its boundaries, and for the same reason does not mention pre-physical inquiries of the kind conducted in *Phys.* I. Lastly, *Phys.* II.1's and VIII.3's lack of any reference to *Phys.* I may have a variety of reasons of little philosophical or chronological significance.

[12] The refutation of the Eleatic thesis that what is is absolutely one occupies part of ch. 2, whereas the resolution of the Eleatic arguments occupies chs. 3 and 8.

reasonable way to tackle issues of this kind is to criticise those who deny them.

Second, those who deny these obvious facts, i.e. the Eleatics, were *forced* to do so by their arguments (*GC* I.8.325a17–18; *Metaph.* A 5.986b27–30) and because they were unable to solve certain difficulties (191b30–1): their claim that change, plurality and nature do not exist (or are not true beings but appearances or false appearances)[13] was not an assumption and a starting point of their reasoning, but rather a conclusion of it.[14] So, the most appropriate strategy to deal with their denial of the starting points of physics is to provide a resolution (*lysis*)[15] of the arguments and the difficulties that brought them to their paradoxical conclusions.

Third, Aristotle presents his solution of the argument that led the Eleatics to deny first coming to be (i.e. the argument that Aristotle calls the '*aporia* of the earliest thinkers') and then plurality altogether as the only effective and correct one (191a23–4). In this he shows himself aware of the fact that he was not the first to tackle the Eleatic theory and try to resist the Eleatics' conclusions by affirming the possibility of natural science.[16] However, Aristotle maintains that only his solution is the correct one. More precisely, he emphasises that only his solution is able to get natural science back on the right path after the Eleatic diversion (191b30–4).

Aristotle's resolution of the *aporia* about coming to be is accomplished in chapter 8 by means of the theory of principles that he constructs progressively in the previous chapters and that he completes in chapter 7. In this discussion, Aristotle does not prove what is more evident by means of what is obscure (193a3–9). Rather, his strategy and goal in chapter 8 may be described as the grounding of what is evident to us (i.e. the *hoti*: the existence of natural, changing things) by means of what is evident in itself (i.e. the *dioti*: the principles of natural, changing things): the first use of Aristotle's theory of principles is to ground the starting points of natural science so as to get natural science back on the right path after the Eleatic diversion.

All this discussion adds further information about the relevance that *Physics* I's treatment of pre-physical issues has for natural science, and so

[13] *Phys.* VIII 3.254a27. [14] Cf. n. 10.

[15] The criticism of the Eleatics' arguments provided in ch. 3 is called *lysis* (I.2.185a8; I.3.186a5, 23). Other passages from *Phys.* I where the terms *lyein* and *lysis* occur are: *Phys.* I.8.191a23-4, 191b30, 34. On the importance of the distinction between *lysis* (resolution) and *elenchos* (refutation), see Rossi 2006, 2014.

[16] On this issue, see e.g. *Phys.* I.9; *Metaph.* N 2.1088b35ff. See also Castelli, Cerami and Lennox in this volume.

about the introductory role of *Physics* I. This book is introductory in a further way because it is aimed at introducing people to Aristotle's own view on natural science by showing how it tackles the ongoing Eleatic challenge to natural science, which influenced the post-Eleatic tradition, how it solves problems that none of Aristotle's post-Eleatic predecessors was able to solve properly,[17] and so by connecting this new enterprise to the tradition of natural philosophy.[18] Moreover, doing this amounts to providing what, in Aristotle's view, seems to be the most reasonable arguments for the starting points of natural science.[19]

To conclude this section: we have collected and analysed some general and macroscopic features of *Physics* I with the aim of clarifying progressively its role and status. This initial survey suggests that *Physics* I has a complex and multi-faceted role. On one hand, it is a scientific physical investigation (indeed, the beginning of Aristotle's science of nature), which shows clear signs of continuity with the other physical inquiries and which introduces principles that would have a foundational role in Aristotle's science of nature and, more generally, in his philosophy as a whole. On the other hand, it also contains some other features that make it look like a relatively free-standing treatise aimed at introducing the science of nature, by setting, as it were, its scene: the book introduces and argues for principles that are not distinctively physical (although they are described as principles of natural things: 190b17–20, and although the aim of the book is to introduce the principle of natural science: 184a14–16); it does so by means of evidence that is not, or at least not always, distinctively physical; it is aimed at tackling in an unprecedented way two basic starting points of natural science, and, by so doing, at putting natural science back on the right path after the Eleatic diversion.[20] To these signs we may add a

[17] For a different view on this issue, see Bodnár in this volume.

[18] In section II, I tackle further the issue about *Phys*. I's target and audience by analysing its various schemes of construction and by emphasising Aristotle's reuse of traditional frameworks.

[19] Aristotle's treatment of the Eleatic theory in *Phys*. I is divided into two distinct parts, which bracket most of the rest of the book: a criticism of Eleatic monism in chs. 2–3 and a criticism of Eleatic immobilism in ch. 8. One aspect of this complex structure is that the treatment of the Eleatic theory is intertwined with the inquiry into the principles. This raises several questions, which clearly complicate the issue of the specific function that the treatment of the Eleatic theory (and of the pre-physical issues concerning the starting point of physics) has for natural science. Why is the treatment of the Eleatic theory divided into two parts? What is the relation between these two parts? What is the relation of each of them with the rest of the inquiry carried out in *Phys*. I? I tackle this issue in Quarantotto (forthcoming), where I suggest that the way in which, in *Phys*. I, Aristotle breaks down the criticism of Eleatic monism and immobilism is functional to the construction of his theory of principles and fits the particular approach of *Phys*. I to this topic.

[20] In section II, I suggest that *Phys*. I is introductory also from a heuristic point of view: it deals with principles that must be tackled first in order to construct a correct theory of principles.

further one that concerns *Physics* I's literary construction. The book is constructed as a relatively completed and autonomous entity: it ends with the claim that its aim has been accomplished and presents what follows it not as a continuation of the inquiry just carried out but as a restart, i.e. a new beginning (192b2–4). Of course, the idea of a beginning followed by another beginning may have a variety of meanings. However, in the light of my initial remarks, a plausible way to understand its occurrence at the end of *Physics* I (on the assumption that these concluding lines are Aristotle's) is that this book has the introductory role of a relatively free-standing treatise. That said, on the other hand, a continuation of *Physics* I's inquiry is needed to accomplish the very aim of *Physics* I: the identification of *distinctively physical* principles of natural things. Moreover, this continuation is already prepared in *Physics* I, especially in chapter 9, where Aristotle distinguishes between the study of form by first philosophy and that by natural science (192a34–b2), and introduces the teleological relation between form and matter (192a16–25), which is a key ingredient in his subsequent inquiry into nature.[21]

In what follows I shall address the issue of the role and significance of *Physics* I by focusing on its overall argumentative strategy, assumptions and relations with other Aristotelian texts.

II. The Argument, its Frameworks and the Status of Aristotle's Principles

The book provides a complex, articulated but unitary argument, whose main conclusion states that there are principles of natural things, what they are and how many they are (192b2–3). This amounts to the well-known Aristotelian theory of matter, form and privation, and to the idea that the principles of natural things are two in one sense (matter – conceived of as that which underlies – and form: 190b17–20, or the two opposites: 190b29–32),[22] and three in another (matter, privation and form: 190b35–191a3).

[21] On this issue, see Lennox in this volume.

[22] These two different ways of counting (form and matter, on one hand, and the two opposites, on the other) do not seem to be on the same level. The first looks like Aristotle's positive achievement, whereas the second seems to be the validation, by means of Aristotle's positive achievement, of the general opinion shared by Aristotle's predecessors that principles are opposites (188a19). Moreover, these two different ways of counting seem to be relative to two different points of view. To count form and matter amounts to focusing both on the principles that are constituents of natural substances conceived of as the result of change (190b20–3) and on the principles from which a substance comes to be (191a4–7). To count the two opposites, on the other hand, amounts either

Physics I is not the only text of Aristotle's corpus where the principles of natural things are dealt with. However, it is the one that provides the longest and most complex argument for the existence, nature and number of the principles that are constituents of natural things:[23] form and matter. If we exclude *Physics* II.3, there are no texts comparable to *Physics* I, from the point of view of its main topic, in the corpus of Aristotle's physical writings,[24] although there are some in his metaphysical writings: *Metaph.* Α 3–10 and Λ 2–5 (besides, of course, *Metaph.* Δ 1–2). *Metaph.* Λ 2–5 provides arguments for the principles. However, in so far as the introduction of form and matter is concerned, it is largely a compendium of *Physics* I: *Metaph.* Λ 2 picks up and summarises some central passages of *Physics* I's argumentation, by skipping its doxographical discussion and by connecting the inquiry to issues that it does not tackle. On the other hand, *Metaph.* Α 3–10 does not argue for Aristotle's theory of principles but presupposes it and tries to make it more plausible (983b5–6) by arguing that none of Aristotle's predecessors had introduced principles that are of a different kind from his own. To do so, it presents and discusses principles that are only imperfect and approximate versions of Aristotle's principles (*Metaph.* Α 7, 10). Further, it is not aimed principally at stressing the differences between these principles and Aristotle's. By contrast, the main purpose of *Physics* I is to argue for a *correct* theory of principles of natural things, i.e. one that, in so far as it is correct, turns out to be different in various respects from the theories of Aristotle's predecessors.[25] For this reason, as we shall see shortly, a long stretch of *Physics* I takes the form of a *diairesis*, or division, whereby the theories of some of Aristotle's predecessors are progressively ruled out.

Further, both *Metaph.* Λ 2 and *Metaph.* Α 3–10 provide clues that, with regard to the distinction between form and matter, *Physics* I's argument has epistemic priority over other Aristotelian texts on this topic. That *Physics* I's argument (i.e. a compendium of it) is used as a means or as an

to focusing on the principles as termini of change or to characterising matter in terms of what it, by itself or before acquiring the form, lacks, i.e. as a composite of matter and privation.

[23] I am here relying on the distinction between internal and external principles drawn in *Metaph.* Λ 4.1070b22–4 and Δ 1.1013a19–20.

[24] *Phys.* II, as a whole, does not focus on the distinction between form and matter as such. Rather, it focuses on the roles of form as mover and as goal, and consequently on the roles of matter as moved-mover and as means. Further, even if *Phys.* II deals with the distinction between matter and form, it does not provide an argument for it. Lastly, *Phys.* II does not tackle privation, and mentions it only once (193b19–21).

[25] For a different view on the relation between *Phys.* I and *Metaph.* Α 3–10, see Mansion 1961: 40; Barney 2012: 73. Both maintain that the inquiry of *Metaph.* Α 3–10 can be seen as in lieu of the argument for the four-cause system that is missing from the *Physics*.

intermediate step within *Metaph.* Λ's inquiry into the first principle of everything suggests that Aristotle views *Physics* I's argument as a good argument and, perhaps, as his main argument for the distinction between substrate, form and privation. Likewise, *Metaph.* A 3.983a33–b6 and A 10.993a10–11 indicate that, in his view, the inquiry into principles and causes conducted in the physical writings is the basis on which *Metaph.* A 3–10's inquiry builds,[26] and that these two inquiries together are sufficient to conclude that the kinds of principles that exist are those and only those he distinguishes.

Physics I argues for a correct theory of principles and seems to have epistemic priority over other Aristotelian texts on this topic.[27] However, it deals with only two of the four Aristotelian kinds of principles: form and matter (with a particular focus on the latter, which is presented as the main achievement of the inquiry (191a15–19, 191b33–6)). The mover and the goal are just hinted at in *Physics* I, and are tackled more extensively in *Physics* II, where Aristotle also provides a long argument for natural teleology. The order in which *Physics* I and II introduce the principles corresponds to that of *Metaph.* Λ 2–5 and, for that matter, of *Phys.* II.3 and *Metaph.* Δ 2. It is different, however, from the order in which the principles are tackled in *Metaph.* A 3–10, which is the following: matter, mover/goal, form. This is a chronological order based on the progressive historical discovery of the principles by Aristotle's predecessors.

Faced with this noticeable difference, one may wonder what the significance of the order followed in *Physics* I–II is. As a first approximation, and in the light of the observations made in the previous section, it seems plausible to hypothesise the following. One of the main aims of *Physics* I is to set the scene for the definition of natural substances that occurs in *Phys.* II.1. Put more carefully, one of the main aims of *Physics* I is to outline an ontological description of natural substances, conceived of as entities that undergo change and that are generable and perishable. This amounts to clarifying what happens in a process of change (in particular, in a process of coming to be), conceived of as a process that unfolds 'from something to something' (*ek tinos eis ti*: 189b32–3), what exactly it is that changes and in what sense. These kinds of questions solicit an inquiry into change and

[26] Since *Metaph.* A 5.986b27–31 mentions *Phys.* I's inquiry into Parmenides' argument for the unity of being, we have a reason to think that the physical writings mentioned at 983a33–b1 include *Phys.* I.

[27] This, however, does not imply that *Phys.* I is the epistemic foundation for Aristotle's subsequent inquiries into principles, i.e. that the account of principles provided there is well grounded and does not have argumentative gaps. On this point, see Charles and Judson in this volume.

into changing things, which combines the progressive distinction of the internal principles that account for the status of natural substances *qua* entities that undergo change, with a particular focus on the substrate/matter.[28] Having done this in *Physics* I, Aristotle proceeds in *Physics* II to inquire into what accounts for the distinction between natural and non-natural substances, and into what the *ousia* and nature of natural substances is. In doing so, he tackles further principles: the principle of change and the goal. Moreover, whereas the focus of *Physics* I is on the substrate/matter as principle from which (*ek*) change occurs and natural substances come to be, as subject of change and as internal constituent of natural substances, the focus of *Physics* II is on form as mover, as goal and as *ousia*/nature, and consequently on matter as moved-mover, as means, and also as nature.[29] So, in a nutshell, the rationale of the sequence followed in *Physics* I–II would be: how a natural substance comes to be from something and what its ontological composition is (i.e. what its internal principles are) (*Phys.* I); what the things that account for its being a natural substance (i.e. for its being in a process of natural change or in a state of natural rest) are, and what the *ousia*/nature of a natural substance is (*Phys.* II).[30]

That said, one may also wonder whether there is a reason for this order of questions and hence for this order of principles. For instance, one may wonder whether the particular heuristics adopted in *Physics* I–II is needed or especially useful to construct a *correct* theory of principles. This would mean that one of the reasons why Aristotle's predecessors did not advance a correct theory of principles, but only succeeded in approaching it, is that they did not follow or did not follow properly or completely the right path and order of research.

There are some clues that suggest this hypothesis. First of all, Aristotle's theory of the two/three principles and of the four causes is a system where each member gets its precise meaning from its role in the whole and from its relations with the other members of the whole. Moreover, it is a system at whose heart there is the distinction between form and matter, for the two other causal roles (mover and goal) are played by the form itself in relation to matter (which is a moved-mover and a means). Accordingly,

[28] From this point of view a useful parallel is offered by *Metaph.* Λ 1–2. There, in the context of an inquiry into substance (1069a18), the notions of form and matter are introduced first of all to define sensible substances *qua* changeable substances (1069b3).

[29] Cf. Philop. *In Phys.* 3.2–6.

[30] Moreover, the inquiry into the nature and *ousia* of natural substances, started in *Phys.* II.1, is closely linked to the inquiry into the teleological character of natural processes that Aristotle focuses on in the subsequent chapters of *Phys.* II (on this connection, see Quarantotto 2005; Kelsey 2015).

one might expect that arguments for this causal system should start from the distinction between form and matter.

This conclusion is also suggested by Aristotle's description in *Physics* I both of the method for discovering the principles and of its use by his predecessors. In *Phys.* I.1. Aristotle describes the method for discovering the principles as a path that has a fixed course and a fixed procedure, and that is natural (184a16) and necessary (184a19, 184a24, 188b29–30), suggesting perhaps that it is correct and that it is in accordance with our human nature and/or with the nature of things themselves. This path goes from what is more knowable to us to what is more knowable by nature, and does so by way of dividing a confounded whole into its parts, and by proceeding from the universal (*katholou*) to the particular (*kath'ekaston*).[31] Further, the fact that the path is described as natural and necessary does not imply that it is easy to follow or that Aristotle's predecessors followed it properly. They employed this method, but they also made several mistakes that prevented them from identifying the right first principles. References to these mistakes are present in crucial passages of *Physics* I. In chapter 1 Aristotle compares the path towards the principles to the ontogenetic development of linguistic abilities in human beings and emphasises a particular linguistic error made by children (184b3–5). In chapter 8 he uses again the metaphor of the path and complains that some of his predecessors (the Eleatics and the post-Eleatics) took a wrong side-track (191a24–7, 191b31–4) because of their inability to solve a puzzle about coming to be. Moreover, he identifies the source of the mistake about coming to be made by the Eleatics and continued by the post-Eleatic philosophers as lying in their failing to see (191b33–4), or failing to see clearly enough (191b35–6), the underlying nature (i.e. Aristotle's concept of matter). This means that they did not distinguish between the substance that comes to be, the underlying nature and the privation.[32] Since this mistake was at the root not only of the Eleatic theory but also of the whole post-Eleatic tradition, it seems that, in Aristotle's view, it is fundamental to constructing a correct theory of principles (and to getting back onto the natural path) that one removes this mistake, by introducing a distinction between the substance that comes to be, its matter and its form.

[31] I take it that the naturalness and necessity of the path attaches not only to its general direction from what is better knowable to us to what is better knowable by nature, but also to the specific route to be followed, i.e. the division of a confounded whole into its parts and the passage from the *katholou* to the *kath'ekaston* (which is said to be necessary). This is suggested also by the presence of a deviation from the path (191a24–7, 191b31–4).

[32] I come back to this point in section III.

Additional support for this hypothesis is found in *Metaph.* A 3–10. As already noted, in these chapters Aristotle describes the progressive discovery of principles by his predecessors. The principles found by Aristotle's predecessors are not his own principles. They are subsumed under his principles because they are seen by Aristotle as tokens of,[33] and approximations to, his principles. What they lack primarily is a satisfactory definition of their relations with the other principles, i.e. of their specific role within the (Aristotelian) causal system as a whole. For instance, the very first philosophers introduced principles that Aristotle does not describe as 'matter' but as 'of the material type' (*en hyles eidei*: 983b7). Moreover, he stresses that matter, as well as the other principles, had been grasped by his predecessors only indistinctly (*amydros*) and in no respect clearly (*outhen saphos*) (985a10–13). This means both that the principles introduced by the first philosophers play the role of matter, and that they play it only approximately and in a confused way.[34] For 'matter' is a relational concept: it is defined in relation to the form and the substance it is matter of. If there is only one principle – i.e. the material principle of the first monists – this can be considered as matter only in a broad and rough sense. From a strict Aristotelian point of view, this unique principle is a confounded whole (in the sense of *Phys* I.1.184a22) that comprises (at least) both matter and form (or both matter and privation).[35] However, the first philosophers did not make the distinction between matter and form. Rather, Aristotle presents the form as the kind of principle that, together with the goal,[36] his predecessors grasped worst, and on which they

[33] The principles introduced by Aristotle's predecessors, with the partial exception of Platonic principles, are not at the same level of generality as Aristotle's principles: unlike Aristotle's principles they are determinate things (hot, cold, odd, even, love, strife, etc.), which Aristotle subsumes under his principles. I come back to this issue shortly.

[34] Note the use of *saphos* at 985a13, which recalls its repeated use in *Phys.* I.1.184a16–26.

[35] Cf. *Metaph.* A 5.986a15–17, 987a13–19 (where Aristotle claims that the Pythagorean number plays both the role of matter and that of affections (*pathe*) and dispositions (*hexeis*), and that these affections and dispositions (e.g. limited and unlimited) were conceived of by the Pythagorics as the essence of what they are predicated of); *Phys.* I.6.189b2–10 (where Aristotle praises those predecessors who tried to distinguish the substrate from its opposite affections); *Metaph.* Γ 2.1004b29–30 (where Aristotle claims that almost all his predecessors agreed that the beings and the *ousia* are constituted out of opposites).

[36] In *Metaph.* A 7 not only the form (988a34–5) but also and especially the goal (988b6–16) are presented as causes that Aristotle's predecessors grasped very little, and at least less than the matter and the mover. Put more carefully, in so far as the goal is concerned, Aristotle claims that some of his predecessors grasped the good, but did so only imperfectly, for they did not use it as a final cause: they grasped a feature of the final cause (its being good), but wrongly attributed it either to the mover or to the one/being; thereby the good turns out to be a cause only accidentally (on this issue, see Quarantotto 2005: ch. 1).

focused last.[37] Tellingly, this criticism of the earlier thinkers for not distinguishing the form is repeated elsewhere, for instance in *Phys.* II.2.194a20–1 and *PA* I.1.642a13–29. So, a possible scenario would be the following: Aristotle's predecessors started by introducing principles that were solely 'of the material type', i.e. whose role was approximately but not distinctively that of matter (a substrate and a subject of change). From this point of view they were following the natural path correctly: they were investigating natural things *qua* changeable things and were inquiring into the status of natural things *qua* subjects of change; moreover, they were trying to tackle this issue correctly by looking for the constituents of natural things.[38] However, one of their first and main mistakes was that they did not articulate this subject/substrate of change into its proper constituents, i.e. into matter and form, but left these in a state of confusion. By so doing, they showed themselves unable to identify the substrate/matter from the point of view of its essential being: the being of the substrate is (i.e. is not distinguished from) the being of its properties (which might be opposites: 189a34–b10).[39] The later philosophers also proved unable to draw this distinction, or to draw it correctly. This failure compromised their theories of principles and left the post-Eleatic tradition of natural philosophy on the wrong path. So, by introducing this distinction in *Physics* I, Aristotle corrects the method of investigation of his predecessors and resumes the correct and natural direction of the inquiry

[37] In so far as the form is concerned, in *Metaph.* A Aristotle mentions only Empedocles (993a17ff.), the Platonists (988a34–b1) and, with less emphasis, the Atomists (985b4–20) as the philosophers who introduced a principle 'of the formal type'. Further, at 988a34–b1 he claims that 'no one has stated clearly the essence and *ousia*, but those who posit the forms speak of it the most'. This remark is consistent with the fact that, in the doxographical inquiry of *Phys.* I, the terminology of form and matter is introduced in connection with the Platonic theory (on this point, see Lennox in this volume). Moreover, the overall discussion of form presented in *Metaph.* A is consistent with that provided in *Phys.* II.1–2. For, in *Phys.* II.1–2, Aristotle claims that among the ancients only Empedocles and Democritus grasped a little the form (194a20–1), whereas the Platonists are described as people who posited forms but wrongly conceived of them as separate entities (193b35–194a1). Lastly, in *GA* IV.1.778b7–10, Aristotle claims that the ancient *physiologoi* did not posit the form and the goal, and that they conceived of the mover and the matter confusedly. This latter point fits with the many places where Aristotle treats earlier natural philosophers, other than Plato, as thinking that everything could be accounted for by virtue of material necessity alone.

[38] On the view of Aristotle's predecessors that principles are constituents of natural things, see e.g. *Metaph.* B 3.998a20–32, Γ 2.1004b29–30 and Judson's interpretation of 188a26–30 in this volume.

[39] In *Phys.* I Aristotle does not say that the material principle of the material monists *is* an opposite. However, the way in which he describes and discusses their theories suggests that, in his view, they lacked a clear grasp of the property/bearer distinction (189a34–b8). Therefore, in Aristotle's view, such theories do not conceive of the material principle as a substrate that is different *in being* from the properties that belong to it.

into nature.[40] Incidentally, this may be taken as a further sign of the introductory and basic character of *Physics* I. The book deals with principles that are basic from a heuristic point of view: they must be tackled first in order to construct a correct theory of principles.

This brings us to the question of the schemes whereby the argument put forward in *Physics* I is organised. As previously said, *Physics* I is a unitary but complex and articulated argument. As we shall see in what follows, its complexity is due, among other things, to its being carried out according to various schemes of organisation, some of which operate together in complete or partial overlap.

One of these schemes is that of the natural path and the progressive solution of *aporiai* along this path, which has been just mentioned. The inquiry conducted in *Physics* I tackles several *aporiai* and the metaphor of the path is used more than once (184a16, 188b26–30, 191a24–7, 191b31–4). Moreover, the discussion of the Platonic theory of principles in chapter 9 is understandable in the light of this scheme. This theory is described as being superficially similar to Aristotle's own, but as not being a correct theory of principles. Further, it is presented as a theory that, unlike Aristotle's, is influenced by the Eleatic approach to the problem of coming to be and does not provide a proper solution of it (191b35–192a3, *Metaph.* N 2.1088b35ff.). One of the reasons why Aristotle discusses the Platonic theory of principles in chapter 9 is to emphasise the difference between his theory and an apparently similar theory that, unlike his own, is unable to solve properly certain traditional problems.

Note, moreover, that the metaphor of the path (i.e. the idea that the acquisition of knowledge is like a path or a journey along a path) is a traditional one, which was introduced and made famous by Parmenides, and developed further by Plato.[41] It is therefore plausible that Aristotle chose to use it in *Physics* I to place his inquiry and achievements in the context of Parmenides' investigation, especially given Parmenides' profound influence on the tradition of natural science.[42] However, this metaphor is used in *Physics* I without the doxographical discussion taking

[40] Moreover, in *Metaph.* A 3 Aristotle stresses that, after positing principles 'of the material type', the ancient philosophers first came to an impasse due to an *aporia* raised by the Eleatics (984a16–b1), and then introduced principles that Aristotle describes as playing approximately both the role of mover and of goal (984b1ff.). If this Eleatic *aporia* and the one mentioned in *Phys.* I are the same or closely connected, this would provide further evidence in favour of the hypothesis here suggested, i.e. that, in Aristotle's view, it is natural and necessary to distinguish between matter and form before positing other principles.

[41] *NE* I.4.1095a30–b4.

[42] I defend and develop this hypothesis in Quarantotto 2016.

the form of a historical narrative. Of course, this does not exclude the underground presence in *Physics* I of a historical narrative and the possibility of reconstructing one out of *Physics* I's indications, but eventually it complicates the understanding of its sense and relevance.

Other evident schemes of organisation employed in *Physics* I are, for instance, the one emphasised in the closing lines of chapter 9, and the division that occurs at the beginning of chapter 2. At the end of chapter 9, Aristotle claims that he has determined that there are principles, what they are and how many they are. This list of achievements can be used to distinguish between chapters 2–3, which deal with the existence of principles,[43] and chapters 4–7, which deal with their nature and number. Further, the division that occurs at the beginning of chapter 2 is a classification of views about the nature and especially the number of the principles. Aristotle distinguishes theories according to which there is one principle from those according to which there is more than one. In the case of the former theories, he distinguishes two ways of understanding the one principle: as something physical or as something that is not physical. In the case of the latter theories, he distinguishes those that posit a finite number of principles from those that posit an unlimited number. This division is used to provide a scheme for the subsequent argumentation, a scheme that overlaps with, but is more fine-grained than, the scheme offered by the closing lines of the book. As such it is better suited to help us uncover the internal workings of Aristotle's reasoning and show the issues at stake, the assumptions and crucial moves.

First of all, Aristotle's argumentation takes the general form of an elimination of the two most extreme views distinguished in the division of chapter 2. These extremes views are associated with theories held by some of Aristotle's predecessors: the radical monism of the Eleatics, which is ruled out in chapters 2–3, and the view that there are infinite principles, whose Anaxagorean version is ruled out in chapter 4. Then, from chapter 5 to chapter 7, the argument takes the form of a positive inquiry aimed at exploring the intermediate view that the principles are many and finite. The purpose of this positive inquiry is to determine the exact finite number of the principles (given that they are more than one but not

[43] In a sense, ch. 4 likewise deals with the existence of principles, for it tackles the idea that each thing comprises an infinite number of principles, which in Aristotle's view cannot be principles of scientific knowledge, i.e. the kind of principles *Phys.* I deals with (187b7–13, 189a11–12).

infinite: 184b19–20).[44] Actually, this positive inquiry also takes partly the form of a progressive elimination of possibilities, but here the elimination is not associated with historically held views. Indeed, unlike in chapters 2–4, in chapters 5–6 the doxographical inquiry is not used negatively to rule out possible views about the number of the principles, but rather as a positive basis to determine the nature and exact number of the principles. In chapter 5, building on a generally held opinion (188a19–30), Aristotle establishes that the principles are at least two opposites (189a9–10); in chapter 6 he argues that they are not more than three (either two opposites or two opposites and one substrate) (189b27–9): he rules out the possibility that there is more than a couple of opposites in the role of first principles and that there is more than one substrate; moreover, he corroborates the hypothesis that the principles are not more than three by pointing to the ancient view that there are three principles, i.e. the one, excess and defect (189b2–16); lastly, in chapter 7 he introduces his own theory of principles (matter, privation and form) and concludes that the principles are in a sense two and in another sense three. Here ends the application of the diairetic scheme.

As noted, the scheme provided by the division of chapter 2 is more fine-grained than the one given at the end of the book. Moreover, its main function is to show how the issue about the number of the principles and its progressive reduction to two/three is dealt with.

Then, there is another version of this diairetic scheme that is used to construct *Physics* I's inquiry. Whereas the first version (that is, the division of chapter 2) considers the principles as such, quite independently from what they are principles of,[45] and focuses on their number, the second version concerns also the relations both between the principles and what they are principles of (i.e. natural substances and their changes) and between the principles themselves. In this way, it offers an additional and important point of view on the argumentative strategy adopted by Aristotle to argue for his theory of principles and on the issues at stake. Indeed, the issue about the relations between the principles and between the principles and what they are principles of is crucial to Aristotle's theory: the very idea that the principles are in a sense two and in another sense three depends upon these relations, i.e. upon the fact that the number of

[44] Note that in ch. 4 Aristotle rules out only the possibility that each thing comprises an infinite number of principles. Instead in ch. 6 he rules out also the possibility that there is an infinite number of principles, which are not the constituents of *each* thing.

[45] In the first version of the division the principles are described as one or many, finite or infinite, changeable or unchangeable (184b15–22).

the principles is considered not only from the point of view of their numerical unity but also from the point of view of their difference in being (190b29–191a3). Accordingly, the issue about the relations between the principles is presented at the end of chapter 7 as one of the main achievements of Aristotle's inquiry (191a18–19).

Moreover, whereas the first version of the division is an 'archaeological' division, which distinguishes between kinds of principles (*archai*), the second version is both an 'archaeological' and an ontological division, which distinguishes between kinds of beings some of which play the role of principles. As we shall see shortly, the shift to a classification that is both 'archaeological' and ontological serves first of all to show that the Eleatic being is not a principle and then to highlight one of the main mistakes made by most of Aristotle's predecessors: the fact that they did not distinguish or did not distinguish properly (i.e. from the point of view of their respective being) between the principles, and between the principles and what they are principles of.

The issue of translating the 'archaeological' division into a division that is both 'archaeological' and ontological is introduced already in chapter 2. At 184b22–5 Aristotle claims that those who inquired into how many fundamental beings exist did in fact inquire into the principles and elements of things, and sketches an ontological version of the first division. The onto-archaeological point of view is then put in use in the critical examination of Eleatic monism, which is described as an ontological theory that posits a unique and changeless being that is not a principle (184b25–185a5). Further, the passage from chapter 3 to chapter 4 makes it clear that this onto-archaeological point of view is used to construct a new version of the initial division, which provides a new scheme of organisation for Aristotle's inquiry.[46] This new version concerns being, conceived of as a unity, different ways in which the relation between being/the one and the many can be intended, and different ways in which change can be conceived of. Indeed, in the passage from chapter 3 to chapter 4 Aristotle first claims that it is impossible that what is is one in the Eleatic way (187a10–11) and then proceeds to consider the physical theories, which are presented as a different way of understanding the unity of what is and its relation to the many (187a12–13, 18–22), a way that in

[46] Furthermore, the employment of the onto-archaeological point of view continues in ch. 6 where Aristotle uses the notion of *ousia*, conceived of as a kind of being (189b23–4); in ch. 7, where he uses the notion of diversity in being to present his own theory of principles (190b33–191a3) and describes his principles both as principles of being and of coming to be (190b17–20); in ch. 8 to formulate and resolve the *aporia* about coming to be.

turn is divided into two further different ways. In this connection, it is particularly striking that theories described as pluralistic in the division of chapter 2 are then presented as forms of monism in chapter 4 (187a20ff.). Indeed, in the initial division provided in chapter 2 (184b15–25) Aristotle distinguishes between a kind of monism that denies change, a kind of monism that admits and accounts for change, and several kinds of pluralism. Instead, in the passage from chapter 3 to chapter 4 these distinctions are reformulated into a distinction between a kind of monism that rules out both plurality and change (i.e. Eleatic absolute monism, which conceives of what is as only one and as changeless) and a kind of monism that admits and accounts for both plurality and change (i.e. physical monism, which conceives of what is as both one and many, and as changeable).[47] This latter monism is itself distinguished into two sub-types.

Arguably, this reformulation is due to the shift in perspective mentioned above: the shift from a perspective that focuses on the principles as such (which is an abstract classificatory perspective) to one that focuses on the relations both between the principles and what they are principles of, and between the principles themselves (which is a heuristic perspective). Indeed, the unitary being that Aristotle attributes to the two physical ways of understanding is indistinctively and confusedly the *terminus a quo* (or, equivalently, the *terminus ad quem*) of change and its constituent principles. Arguably, this being is conceived of as unitary (and hence the physical theories are described as forms of monism) since this is how natural substances appear at the beginning of the inquiry (i.e. as unitary wholes: 184a21–2, 24–5) and since unity is one of the basic conditions that something must satisfy to be a subject of change. Indeed, the numerical unity of substances will be one of the central tenets of Aristotle's theory of principles in chapter 7. However, in order for a unitary being or substance to undergo change it must also be articulated into parts, i.e. it must be not only one but both one and many.[48] If it were only one, i.e. an

[47] The second physical way of understanding (in particular Empedocles' and Anaxagoras' theories) is characterised by the fact that beings are conceived of as both one and many (187a21–2). Instead, in so far as the first physical way of understanding is concerned, the fact that it views being as both one and many, and hence that it is a kind of monism that admits plurality, stems from the fact that the unitary being, which plays the role of principle, is conceived of as a confounded whole comprising both substrate and opposite (i.e. without a distinction in being between a property and its bearer). This is not emphasised immediately in the division that opens ch. 4 (187a12–26), but it is stressed soon afterwards (cf. 189b2–16 (on the Ionians), 192a1–12 (on the Platonists), and 190b30–191a3, where Aristotle presents the distinction in being between substrate and opposites as his own achievement).

[48] The principles of natural things – which Aristotle presents as the final result of the correct application of the method prescribed in ch. 1 – are constituents (*stoicheia*) of natural substances

absolute unity like the Eleatic being, it would be changeless. Accordingly, the physical theories are described as theories that posit a unitary being that is articulated into parts.

Hence, this second, onto-archaeological, version of the division can be seen as a reformulation of the first version, aimed at tackling the relation between one and many (i.e. between a natural substance and its internal principles), and the way in which something unitary can be internally articulated, which in Aristotle's view is one of the basic conditions for change. On this reading, from chapter 2 onwards Aristotle inquires into the ontological structure that enables something to undergo change. In particular, in chapters 2–3 he rules out the Eleatic hypothesis about being/the one, which is incompatible with change (i.e. the view that what is is absolutely one and hence changeless), whereas from chapter 4 onwards he considers the physical hypotheses, which are instead compatible with change (i.e. the views that what is is both one and many, and hence subject to change). The inquiry conducted from chapter 4 onwards is aimed at determining progressively the nature, number and relations of the many that are constituents and principles of the one, i.e. at articulating into its first principles the unity both of that from which change occurs (i.e. the *terminus a quo*, whose constituent principles, in Aristotle's view, are matter and privation) and of natural substances conceived as the result of change (i.e. the *terminus ad quem*, whose constituent principles, in Aristotle's view, are matter and form).

This is done first of all by examining the physical theories of Aristotle's predecessors. The main difference between the two physical ways of understanding concerns the manner in which they conceive of both the relation between one and many and the kind of change that the one brings to the many. According to the first way of understanding, being/the one is a substrate from which the other things are generated by means of a pair of opposites in the role of forms and differences. Thus, change is conceived of as the modification of a substrate. Instead, according to the second way of understanding, being/the one is a unity that comprises within itself a plurality of oppositions. Hence, change (i.e. the process whereby the many different things are generated) is conceived of as a

(189b16). Moreover, the internal articulation of natural beings into parts that are their principles is presented as a central aspect of the ontological structure whereby natural beings are subjected to change (190b11, 20–7). On this issue, see Quarantotto (forthcoming), where I argue that, in Aristotle's view, the existence of principles of natural things requires a special kind of pluralistic ontology, i.e. one in which being is both one (the whole) and many (the parts).

separation from being/the one, which in Aristotle's view is not properly a process of coming to be (187b23).

After ruling out the Eleatic option in chapters 2–3, the argumentation proceeds by excluding the Anaxagorean version of the second physical way of understanding (in chapter 4). By so doing, Aristotle excludes a way of understanding that does not posit a substrate or anything that is close to being a substrate (in the sense of what underlies).[49] Instead, the first version of the physical way is not ruled out, or at least not in the radical manner in which the other ways are. Rather, after excluding in chapter 4 the Anaxagorean version of the second physical way of understanding, in chapter 5 Aristotle starts a positive inquiry that will lead to a result that is an intermediate version of Aristotle's own theory and that is similar, in some relevant senses, to the first physical way (189a34–b16).[50] Lastly, in chapter 9 the Platonic version of this first physical way is presented as a theory that displays some relevant similarities with Aristotle's own.[51] This suggests that Aristotle sees the first version of the physical way as the best approximation, among the theories of principles formulated by his predecessors, to the correct theory of principles, and that he is employing the doxographical investigation to move towards his own theory of principles and especially to its cornerstone, i.e. the notion of substrate.[52]

[49] In *Phys.* I Aristotle does not use the term *hypokeimenon* (in the sense of what underlies change or of subject of predication) to describe Anaxagoras' theory of principles, except to criticise it (188a7–13). Moreover, in *Phys.* I.6 he does not use the theories of the second group of natural philosophers singled out at the start of I.4 to provide support for the introduction of a third principle that underlies the opposites. Arguably, this is because a mixture, as such, cannot play the role of substrate of what it is a mixture of, since all beings (apart from the *nous*) are ingredients of the mixture. So, they are not really distinct from the mixture and cannot be the things of which the mixture is the substrate. Further, although Anaxagoras posits principles that are not opposites, i.e. the uniform ingredients in the mixture, he does not use them as substrates for opposites. Instead, Anaxagoras' theory does posit a substrate in the sense of *terminus a quo* of change (i.e. what pre-exists): see 187a26–31. For a different point of view on the role of Anaxagoras' mixture, see *Metaph.* A 8.989a30–b21, where Anaxagoras' theory of principles is assimilated to the Platonic theory of the One and the Different or Indefinite, and *Metaph.* Λ 2 1069b20ff., where Anaxagoras is praised since his notion of mixture comes very close to Aristotle's notion of matter, conceived of as what is potentially both opposites.

[50] Both in ch. 4 (187a12–13, 19) and in ch. 6 (189b5–7) the principles of the first way of understanding are described as *hypokeimenon* (in the sense of what underlies), on one hand, and as forms and differences, on the other. Moreover, in ch. 6 (189b1–16) Aristotle mentions a general version of the first physical way of understanding (the one, excess and defect: 187a16–20, 189b10–13) to corroborate the preliminary results of the positive inquiry conducted from ch. 5 onwards.

[51] Plato came close (or closer than any other philosopher before Aristotle) to identifying the substrate and to distinguishing it from the form (191b35–6).

[52] The contraposition between the Eleatic view on being/the one and the first physical way of understanding being/the one, which is described as a substrate (187a13, 18–19), suggests that Aristotle's focus here is on the fact that the Eleatic being/the one is not a substrate.

But Aristotle's theory of principles is clearly not identical with any theory included in the first physical way of understanding. Moreover, it is not argumentatively built on the basis of this way of understanding, but is arrived at by following a research path in which the first physical way of understanding occurs only at an intermediate point to corroborate some results. So, it is important to analyse this path in order to understand how Aristotle builds his theory and differentiates it from that of the first physical way of understanding.

The first physical way of understanding includes both the Ionian theories of principles and the Platonic theory of the dyad and the one. Neither Ionian nor Platonic philosophers grasped completely the substrate and the correct relations between the principles. The Ionian philosophers did not grasp the difference in being/account between the opposites and the substrate (189b2–16). Plato approached the notion of substrate and the distinction between substrate and form, but wrongly identified the substrate with what is not, without grasping that the substrate is a composite entity comprising a matter-component and a privation-component (192a1–12). Moreover, none of these thinkers grasped the true *first* principles. Aristotle distinguishes the principles of his predecessors on the basis of their being either universal (*katholou*) or particular (*kath'e-kaston*), and seems to maintain that the universal principles are better than the particular ones (188b35–189a9). However, as we shall see shortly, the reason why these principles are not true *first* principles is not simply that they are not universal or not sufficiently universal, but that they do not possess the right kind of universality. Moreover, as we shall see shortly, there seems to be an intimate connection between the issue about the relations between the principles and the issue about the kind of universality that belongs to them.

The difference between Aristotle's theory of 'matter, privation and form' and the theory of 'the one, excess and defect' (the general version of the first physical way of understanding) is built in chapters 5–7. In order to understand this difference and the epistemic status of Aristotle's theory, we must understand the argumentative strategy that leads to it.

As already stressed, from chapter 5 onwards Aristotle's inquiry takes a different course from the previous chapters. It does not proceed by considering historically held views and by eliminating some of these, but (among other things) uses general opinions (*endoxa*) to conduct a positive investigation. The argumentation proceeds by introducing first the opposites and then the substrate. The opposites are introduced by means of a very general *endoxon*: all the previous thinkers made the opposites principles (188a19). Arguably, one of the reasons why Aristotle starts from an *endoxon* about the opposites rather than from an *endoxon* about the substrate is that the principles introduced by his predecessors (with the partial exception of the Platonic principles dealt with in chapter 9) are opposites, i.e. at the very least there is no clear distinction between the being of the opposites and the being of the substrate.[53] Put differently, Aristotle's predecessors had arguments for the opposites (188a27–30), but lacked arguments for the substrate as such (i.e. as distinct in being from its properties and hence from the opposites). The general *endoxon* about the opposites, by itself, only serves to establish that the opposites are principles, but it is not sufficient to determine the specific role played by the opposites, how many opposites are principles, which opposites are principles, if not all of them, and if besides the opposites there are other kinds of principles. The subsequent argumentation in chapters 5–6 is aimed at sketching the role played by the opposites (i.e. as constituents and as termini of change: 188a27–30, 188a30–188b26), at reducing the number of the opposites that play the role of principles (189a12–20, 189b16–27), at introducing a third principle that plays a different role from the opposites (189a21–189b16), and at reducing the number of items, which play this further role, to one (189b18–19).

These chapters put the positive basis for the theory of principles that Aristotle completes and formulates in chapter 7. The notions introduced and defended prior to chapter 7 correspond to an intermediate phase of the construction of Aristotle's theory. The notion of substrate is introduced on the assumption that if something changes it must be somehow affected,[54] and by means of a very general functional distinction: the opposites are affections and the substrate is what is affected by them (189a22–6).[55]

The next step is to add that this further element – whose role is to be affected by the opposites – has a different being from the opposites, is a principle and has priority over the opposites. To reach this conclusion, Aristotle employs the notion of *ousia*, conceived of as subject of predication, which is different both in account and functionally (i.e. from the

[53] See 188a20–2 (where the principles that Parmenides calls fire and earth are respectively (the) hot and (the) cold), 189b1–8, 192a1–12. On this issue, see n. 47.

[54] Prior to ch. 6 the notion of substrate, in the sense of subject of predication, was already tackled in chs. 2 (185a31–2) and 3 (186a29, 34).

[55] Broadly, the argument goes as follows: if there are opposites, and if change implies that something is affected, then the opposites must be affections and there must be something that is different from the opposites and that is affected by the opposites, for opposites cannot be affected by each other (see Code in this volume).

point of view of its role) from what is predicated of it, and which is what the predicates depend upon (189a29–34). In sum, the substrate is here conceived of as what is affected by the opposites and as *ousia* in the sense of, or on the model of, subject of predication. It will take further steps to arrive at the final physical notion of substrate. However, these intermediate steps show that Aristotle is here putting the inquiry on a level of great generality, which is a functional kind of generality (for it concerns the different roles needed for there to be change), and the generality of *ousia*, here conceived of as subject of predication (which is again a functional concept)[56] and as the most general description of natural entities *qua* entities that undergo change. Indeed, at 189b23–4 *ousia* is described as a kind of being. Moreover, the most proper feature of *ousia* is that it is receptive of opposites while remaining numerically one and the same during change (*Cat.* 5.4a10–11). So, this is the level of generality where Aristotle finds the tools to start distinguishing between opposites and substrate (i.e. to start disentangling the confounded – e.g. Ionian – whole that comprises opposites and substrate), and to posit principles that are common to all things that change, come to be and pass away.

The level of *ousia* is used again to reduce the number of opposites to two primary opposites (189a12–20, 189b16–27). Here, as already stressed, *ousia* is described more precisely as a single kind of being (189b23–4). Broadly, the idea is that in each kind there is only one pair of primary opposites and that this holds also for the kind of *ousia*, conceived of as a kind of being.[57] The implicit assumption is that the principles sought for are the opposites that belong to the kind of *ousia*, which is the most general kind to which natural things belong. In chapter 6, Aristotle does not yet identify the primary opposites, and does not introduce the notions of form and privation. However, he stresses the need to reduce all the opposites to a single pair of primary opposites that are principles and elements of all things (189b26–7). From *Phys.* I.7 we learn that these primary opposites are form and privation. Moreover, from *Metaph.* Λ 4 we learn that the reduction to the primary opposites is of an analogical type (1070a31–3, see also *Phys.* I.5.189a1): the primary opposites are not particular entities or kinds of particular entities; rather, they are analogical and functional entities, which pick out the role played in change by all opposites.[58]

[56] Thereby I mean that the distinction between substance and accidents concerns also their different predicative *roles*.

[57] On this point, see Code in this volume.

[58] However, the form that is presented as a principle of the being and of the coming to be of natural substances in ch. 7 (190b17–20) is a substantial form, i.e. an analogical entity ranging over various

The same holds for the notion of matter or underlying nature, which in chapter 7 is said to be knowable by analogy (191a7–12).

On the kind of universality of Aristotle's principles some interesting insight is offered by *Metaph.* I 1.1052b1–15. Here Aristotle distinguishes between what is an element and the essence of element. He claims that fire is an element (i.e. 'element' is predicated of fire), but the essence of element and the essence of fire are different. 'Element' is predicated of fire because the name 'element' signifies that a certain feature (i.e. a certain role in something: 'there is something which is made of it as a primary constituent': 1052b14) belongs to something, i.e. to fire.

In the light of this distinction, we are able to state that the principles argued for by Aristotle in *Physics* I are the essences of the things that play the role of principles as such: they are the *ti esti* of principles *qua* principles, and this is why they are *first* principles.[59] These essences are functional and analogical beings, which are distinct from the being of the various things that, in the different actual circumstances, play the roles that these first principles pick out. From this point of view, Aristotle's first principles do not exclude other things from being principles in the various actual circumstances. Quite the contrary, Aristotle's first principles are *what account for* the fact that other things play, in the various actual circumstances, the role of principles, i.e. of proximate or proper principles, and of particular principles. They only exclude other things from being *first* principles.[60] In this sense, the functional and analogical universality of Aristotle's first principles, i.e. what makes them *first* principles, is intimately connected with the procedures whereby these principles are distinguished from each other, for this distinction is a functional one: it concerns their respective roles, i.e. the definition of their respective essences in functional terms.[61]

kinds of substances, not over various categories. But the notion of substantial form, although perhaps hinted at and required by the idea that a substance cannot be from non-substances (189a33–4), is not tackled in *Phys.* I (191a19–20).

59 In *Phys.* I.5 (188b30–189a9) Aristotle distinguishes between principles that are more knowable by perception and principles that are more knowable in account. Moreover, he seems to consider the latter to be better than the former, and hence to imply that the true first principles are those principles that are more knowable in account. If this is correct, it would support the hypothesis just advanced, i.e. that form and matter are first principles in so far as they are the *ti esti* of principles as such (since the *ti esti* is expressed by the definitional account and is therefore the 'most knowable in account').

60 In this sense Aristotle's primary opposites are the principles of the *systoichiai* of opposites (on Aristotle's uses of *systoichiai* and especially on his use of it in *Phys.* I.5, see Lefebvre 2008).

61 At 191a7–8 Aristotle claims that the underlying nature is knowable by analogy. Then, at 192a31–2 he says that matter is 'the primary underlying subject for each thing, from which, as a constituent,

So, in *Physics* I, Aristotle is following the method of *APo* II: he first asks if there are principles (*ei esti*), and then what the principles are (*ti esti*).[62] Further, the question 'what the principles are' is intertwined with the question 'how many they are and in what sense', because the reply to the former question requires dividing a confounded whole (a numerically unitary substance) into its functional parts (into the different functional beings it comprises). Aristotle's reply to the second question ('what the principles are'), and in particular the fact that this reply is given on the level of essence, is what chiefly shows the distance of his theory of principles from the theories of his predecessors (with the partial exception of the Platonists).

All this highlights some of the advantages of Aristotle's approach over his predecessors' (besides the solution of the *aporia* about coming to be, stressed in chapter 8). Aristotle's approach leads to the identification of principles that (a) unlike the particular principles of some of his predecessors (e.g. the rare and the dense, the hot and the cold) are able to account for all natural things and do not undermine the existence of proper or proximate principles, and that (b) unlike the more universal principles of other predecessors (e.g. the great and the small) do not undermine the existence of particular principles.

However, the issue of the kind of universality of the principles is only hinted at in *Physics* I, and not made explicit, at least not in the emphatic way in which this is done in *Metaph.* Λ. Moreover, the focus of *Physics* I is not on all natural things and processes but on those things that come to be and pass away, and on finite processes that occur between opposites. The issue of the principles of natural eternal things is not even addressed. Arguably, this exclusion is likewise explained by Aristotle's method that requires starting from what is better knowable to us (i.e. corruptible things and finite processes).

something comes to be non-incidentally', which is arguably a definition of matter conceived of as an analogical unity. Instead, in so far as form is concerned, there seems to be no attempt at defining it. The form is described as a *kategoroumenon* (something predicated of something else) at 190b29, and as 'that of which the account [*logos*] is' at 191a13. This asymmetry may be due to the fact that the focus of *Phys.* I is on the substrate/matter, and to the fact that the question 'what the form of natural things is' can be answered properly only after, or together with, an answer to the question 'whether the form or the substrate is the *ousia*' (191a19–20), a question that Aristotle tackles in *Phys.* II.

62 However, from the terminological point of view, in *Phys.* I there is no explicit talk of the *ti esti* question. At 191a20–2 Aristotle's discourse is about *posai* (how many), *pos posai* (in what sense how many), *tines* (what) the principles are, and about their *tropos* (the way in which they are principles). At 192b2–3, he claims that he has established that there are principles (*hoti*), what they are (*tines*) and how many they are (*posai*).

III. Background

To clarify the role of *Physics* I and the kind of introduction it provides to Aristotle's physical writings it is also useful to understand its relation with other treatises of the corpus. In particular, this section focuses on some of the main assumptions of *Physics* I's argument, and asks whether it depends on other works by Aristotle. The focus is primarily on the relation between *Physics* I and the group of treatises that came to be known as the *Organon*.

There are several reasons why this latter issue is particularly relevant here. First, two works of the *Organon* – the *Topics* and the *Analytics* – focus especially on first principles. The *Analytics* gives an account of the nature of first principles and tells us how to deploy them in scientific demonstration. The *Topics* is, among other things, meant to help us develop the 'philosophical sciences', especially their first principles (I.2.101a26–7, a34–b4). However, neither of them mentions form and matter, the first principles introduced in *Physics* I.[63] A possible reason is that the *Analytics* and the *Topics* tackle the principles in a way that is intended to be valid for any branch of theoretical knowledge and for any philosophical discipline respectively.[64] On this view, they may be considered preliminary to the inquiries into specific kinds of principles. In this connection, part of the tradition, at least since Andronicus, maintains that the writings that comprise the *Organon* are designed to prepare the student for the scientific and ethical works that come later in the order of learning, and this view is shared also by several recent interpreters.[65] Secondly, in so far as principles are concerned, the *Categories* contains hardly any talk of principles and causes.[66] In particular, like the rest of the *Organon*, it does not mention form and matter. However, the *Categories* introduces a concept of *hypokeimenon* that is used several times in the argument of *Physics* I, and that has some relevant features in common with the concept of *hypokeimenon* outlined in *Physics* I. More generally, in *Physics* I, right from the start of the

[63] Put more carefully, they do not use the terminology of 'form' and 'matter'. See for instance *APo* II.11, where – like e.g. in *Phys.* II.7.198b5–6 – Aristotle speaks of the material cause as 'that if certain things hold it is necessary that this does', and of essence instead of form. Of course, even if what these treatises lack is only the terminology of form and matter rather than the respective concepts, this does not imply that, there, Aristotle uses *his* concepts of form and matter.

[64] See Burnyeat 2001: 98. Of course, this hypothesis may be true even if the *Analytics* and the *Topics* employ Aristotle's own concepts of form and matter. That said, my point here is not whether the *Organon* uses or presupposes the concepts of form and matter, but whether it contains background knowledge that is a necessary prerequisite for *Phys.* I's argument.

[65] This view can be suggested also by texts such as *Metaph.* Γ 3.1005b1–4.

[66] The exceptions are chs. 12–13 on priority and simultaneity.

inquiry, we meet terms, concepts, conceptual systems and distinctions that also occur in the *Organon* and that are employed there in a way that is arguably standard or basic: for instance, the distinction between the categories, the notion of privation, the distinction between universal (*katholou*) and particular (*kath'ekaston*), the notion of science (*episteme*).

So, on this basis, one may raise the following questions: Does *Physics* I presuppose the *Organon* or part of it? And if so, in what way? Is the argument of *Physics* I based on assumptions that are justified in the *Organon*? Is it based on data that are provided there? More generally, is *Physics* I's argument based on distinctively Aristotelian assumptions or data that are respectively grounded or provided elsewhere in the corpus and whose identification may help us clarify the role of *Physics* I and the epistemic status of its account of principles?

In what follows I shall focus on the way in which terms and notions used in the *Organon* are employed in *Physics* I. I shall try to clarify what function they play in the argument of *Physics* I and what their epistemic status is.

Let us start from the beginning of *Physics* I's inquiry. The first sentence of the book grounds the inquiry to be developed on an assumption about the nature of scientific knowledge, which is defined as the knowledge of something in terms of its first causes, principles or elements (184a12–14): Aristotle applies this definition to the study of nature and draws the conclusion that the search for principles is the first task of natural science (184a14–16). Then, the subsequent passage describes the method whereby the principles become known to us: it consists in following a certain path that proceeds from what is clearer and more knowable to us to what is clearer and more knowable by nature, from a confounded whole to its parts, and from what is universal to what is particular (184a16–26).

This beginning contains various clear references to topics tackled in *APo*: the definition of scientific knowledge (71b9–12), the distinction between principles and things known through their appropriate principles – a distinction drawn in terms of things more knowable by nature and things more knowable to us (71b29–72a5) – the description of the path towards the principles (II.19),[67] and also, at least implicitly, the conception of physics as a particular science within a broader scientific domain, which is stressed again in chapter 2 with particular emphasis (185a1–3, 14–17), and which occurs lastly in chapter 9 (192a34–6). These multiple references, also because of their cumulative effect, might suggest that the

[67] See Bolton 1991.

inquiry conducted in *Physics* I relies on Aristotle's own conception of scientific knowledge, of natural science, of principles, causes and elements, and of the method for discovering principles, causes and elements. However, as we shall see shortly, this is far from being obvious. Indeed, the way these topics are referred to does not impose such a conclusion. Rather, there are reasons to think that Aristotle's own views on these topics do not provide the starting point of the main argument of the book (i.e. the argument aimed at 'determin[ing] ... the things concerning the principles').[68]

First, the notion of scientific knowledge. The notion of scientific knowledge mentioned in *Phys.* I.1 is clearly connected with the definition of *episteme* provided in *APo.* However, there are not binding reasons to think that Aristotle is here presupposing his own theory.[69] What Aristotle presupposes here is more likely a shared view that has nothing specifically Aristotelian. Indeed, the idea that there is a special kind of knowledge, which is based on principles, causes or elements, was shared also by Aristotle's predecessors. This is a fact that Aristotle stresses several times in his writings (e.g. *Metaph.* A 3.983a24–b4, A 1.981b25–982a2; *Phys.* I.5.188b28). More tellingly, this is a fact that is arguably meant to provide the basis for the doxographical inquiry conducted in *Physics* I: the discussion of the theories of principles of Aristotle's predecessors. Indeed, with the sole – and partial – exception of the Eleatics, these predecessors are described as natural scientists,[70] i.e. as people engaged in a scientific inquiry into natural things (187a12). Further, if the tradition of natural philosophy is at least part of *Physics* I's target and audience, as I suggested earlier, the hypothesis that Aristotle is here appealing to a common assumption of this tradition gains additional plausibility.

A similar account seems to hold also for the notions of principle, cause and element: *Physics* I does not seem to presuppose a distinctively Aristotelian view on what a principle, a cause or an element is.[71] Rather, the

[68] In this section I emphasise reasons that relate especially to *Phys.* I and that are additional to those stemming from the hypothesis that the *Analytics*' view on principles and on demonstrative science is content-neutral and non-partisan (on this hypothesis, see Burnyeat 2001: 99–111).

[69] The use of the plural at 184a12–14 does not imply it (see Falcon in this volume).

[70] On the Eleatics, see n. 10.

[71] Naturally, the distinction *between* principles, causes and elements, mentioned at 184a11–14, was not common or standard. However, this terminological distinction is not important for the inquiry conducted in *Phys.* I (see Wieland 1992: 69): during this inquiry Aristotle uses all the three terms; although 'principle' and 'element' are more frequent than 'cause', the word 'cause' occurs, together with 'principle', to describe Aristotle's own theory as a theory of 'causes and principles' of natural things (190b17). Of course, there remains the possibility that the distinction between principles, causes and elements has a proleptic function for subsequent inquiries.

book is aimed at *discovering* the principles of natural things, and does so by means of a combined method that comprises not only an inferential procedure from effect to cause (i.e. from change and changing things to their principles), but also the discussion of Aristotle's predecessors' theories of principles (which will turn out to be unlike Aristotle's own, in some relevant respects). The entire inquiry seems to be based on very few and rather common presuppositions about principles: basically, (1) the idea that principles are starting points both of scientific knowledge and of the things they are principles of; (2) the idea that at least some principles are constituents of what they are principles of; and (3) the idea that these constituents are principles both of the being and of the coming to be of natural things (190b17–20). Of course, the issue whether Aristotle's predecessors correctly identified the principles is a different matter. Further, the claim that they did not identify the true first principles is not the starting point of Aristotle's inquiry in *Physics* I but part of its conclusion.

(1) The idea that principles are starting points both of knowledge and of the thing they are principles of occurs in various passages of *Physics* I: in chapter 1 (184a10–14, 184a21–3), 2 (184b23), 6 (188a27–8), 7 (190b17–20) and *passim*. This meaning is very close to the one presented in *Metaph.* Δ 1.1013a17–19 as the common feature shared by various senses of 'principle': 'it is common to all to be the first point from which a thing either is or comes to be or is known'. Thus Aristotle seems to be drawing on common views about principles, and to be summarising a shared idea about what a principle is (as it is suggested also by *Phys.* I.5.188a27–30): a principle (*arche*) is a beginning, i.e. something which is prior (in some sense of 'prior') to other things, which provides some kind of foundation for other things (185a4–5, 188a28), and which is not preceded by anything else (188a28, 189a30–2), i.e. which remains (189a19–20). So, in so far as Aristotle is appealing to a shared view about principles of natural things and of scientific knowledge, and in so far as, in *Physics* I, he criticises both the Eleatic and the Anaxagorean challenges to this view, this view of the nature of principles is at least dialectically justified. Indeed, after the criticisms of the Eleatics' and Anaxagoras' theories provided in chapters 2–4, from chapter 5 Aristotle starts building his own theory of principles on a thesis about *primary* opposites (188a28–29) that is presented as a general opinion.[72] Further,

[72] Incidentally, this point of view may help explain why, although both Anaxagoras and Democritus held that there are infinite principles, Anaxagoras' theory is the target of a severe criticism, whereas Democritus' theory is listed among those who illustrate the general opinion from which Aristotle

while constructing his own theory of *first* principles, Aristotle provides reasons for their being first, in the specific sense in which Aristotle's principles are first: (i) in chapter 5, when he reduces the opposites to a single pair of *primary* opposites in the kind of substance; (ii) in chapter 9, when he argues that his notion of matter does not involve an infinite regress, i.e. that matter is, in a sense, ungenerable and imperishable (192a25–34). Above all, the argument of *Physics* I as a whole provides reasons for the need of there being *first* principles in the specific sense in which Aristotle's principles are first.[73] Of course, there are other senses in which principles may be infinite and in which an infinite regress of principles may occur (thereby challenging the notion of *first* principles), but these are not tackled in *Physics* I, as they are not relevant to its main argument.

(2) The hypothesis that *Physics* I focuses on principles in the sense of constituents (*stoicheia*),[74] and assumes that at least some principles are constituents, is suggested by several passages of the book. In chapter 1 principles are presented not simply as starting points of the thing they are principles of, but more specifically as the result of a division of a whole into parts (184a21–3). If this whole–parts image is not so loose as to apply to any kind of principle (i.e. to external principles as well), this suggests that principles are here conceived of as constituents (of some kind) of what they are principles of. Further, the idea that principles are constituents seems to be assumed also in the arguments against the hypothesis of infinite principles put forward in chapters 4 (187b7–13) and 6 (189a12–13). Lastly, the same assumption seems to be implicit in chapter 5 (188a27–30), where Aristotle explains why his predecessors' view that principles are opposites is reasonable.[75] Now, the assumption that at least some principles are constituents seems to be grounded, in *Physics* I, both dialectically and scientifically. It is grounded dialectically, since the conception of principles as constituents was standard among Aristotle's predecessors (*Metaph.* A 3.983b6–11). It is grounded scientifically, since Aristotle describes the procedure whereby these constituent principles become known to us as a natural (184a16: *pephyke*)[76] and necessary path (*anagke*: 184a19; *dei*: 184a24). Arguably, these two aspects are connected

starts his positive inquiry: unlike Democritus' theory (because of Democritus' distinction between different levels of opposites: kinds and instances of kinds), Anaxagoras' theory is incompatible with the notion of '*first* principles'.

[73] I develop further this point in section II.

[74] *Stoicheia* in the sense given in *Metaph.* Λ 4.1070b22–4.

[75] On this point, see Judson in this volume. [76] On this point, see Falcon in this volume.

in Aristotle's view: the conception of principles as constituents was standard among Aristotle's predecessors, since their physical inquiry corresponds to the first segment of the path that leads to the discovery of principles, i.e. the segment that begins with principles that are constituents. That said, however, one may still wonder what grounds Aristotle's view that the search for principles of natural things starts naturally from such principles.

(3) The idea that these constituents are principles both of the being and of the coming to be of natural things is clearly expressed in chapter 7 (190b17–20).[77] One may wonder what the exact meaning of this idea is. Clearly the idea that the principles of natural things are principles both of their being and of their coming to be is consistent with the way in which the inquiry into principles is set up: it is the search for principles of entities that undergo change (185a12–13) and whose being and existence are the result of change (192b1). From this point of view, this idea seems to be an assumption shared by Aristotle's predecessors, as is shown by the fact that their principles are presented both as principles of the being (e.g. 184b22–5) and as principles of the coming to be (e.g. 187a12–26, 188b34–5) of natural entities. So, at least dialectically, this idea seems to be sufficiently grounded in *Physics* I. This, of course, does not imply that it is also scientifically grounded, and that in *Physics* I Aristotle provides all the tools needed to understand and justify it fully. I shall come back to this latter issue in the next section.[78]

If this is correct, Aristotle does not presuppose, in *Physics* I, a particular conception of principles of his own. The same seems to hold of his method for discovering the principles: it is not presented as a particular method of Aristotle's own, but, as already stressed, as a natural and necessary one: it is seen as a procedure that all human beings are naturally inclined to follow and that therefore also Aristotle's predecessors did follow, at least in part.

Let us now move to the conception of physics as a particular science within a broader scientific domain. As already recalled, this conception occurs both in the opening lines of chapter 1 and again, and more explicitly, in chapter 2, where Aristotle employs the thesis that the scientific domain is divided into parts, each of which has its own start-up assumptions and is not expected to justify them (185a1–3, 14–17). Moreover, the conception of physics as a particular science is referred to

[77] By 'principles of coming to be' I mean 'the principles *of natural substances* that account for their coming into being' (see e.g. Wieland 1992: 110; Giardina 2002: 24) and not 'the principles *of change*' (as some commentators have held; for some references, see Judson in this volume; see also Sorabji 2012: 2).

[78] On this issue, see also both Charles and Judson in this volume.

also in chapter 9, where Aristotle mentions the distinction between physics and first philosophy (192a35–6). This is a view that we are familiar with also from several other Aristotelian texts, but which was not common among Aristotle's predecessors. So, these passages show that Aristotle is here relying on a conception of natural science, and of science in general, that was not a traditional and standard one. However, this peculiar conception of Aristotle's does not play a relevant role within the inquiry into principles carried out in *Physics* I. As already emphasised, in *Physics* I Aristotle does not even provide a definition of 'nature' and of 'natural thing', i.e. of the central objects of physics conceived of as a particular science. Rather, he starts from a very broad description of how the central object of natural science (which was also the central object of the inquiries of his predecessors) appears at the beginning of the investigation: as what is more 'knowable to us' (184a16–17), as what is universal and perceptible (184a23–6), and as a multitude of things some of which undergo change (185a12–14). So, although *Physics* I contains references to a conception of physics as a particular science and to the distinction between physics and first philosophy, these are not included among the assumptions needed to conduct the inquiry into the principles of natural things.

In sum, the considerations advanced so far suggest that, in *Physics* I, Aristotle relies on assumptions and views on scientific knowledge, principles and method that he considers to be common to and shared by his predecessors. Further, besides being dialectically justified, most of these assumptions receive in *Physics* I a further – scientific – justification. It remains to be seen whether – when a scientific justification is not provided – a dialectical justification is sufficient to ensure *Physics* I's argument and main achievements a solid epistemic status.

Let us move to the *Categories*. The role of the *Categories* in *Physics* I seems, at least at first sight, especially relevant. Several times in *Physics* I Aristotle employs concepts and distinctions that are tackled in the *Categories* and that are traditionally considered as belonging to the *Categories*: in particular, the distinction between the categories – especially the distinction between substance and accidents – the idea that accidents depend upon substances, and the distinction between essential and accidental predication. In *Physics* I these concepts and distinctions are among the main tools used to get to the distinction between matter and form, and are employed in various crucial steps of the argument: mainly in chapters 2, 3, 6 and 7.

Let us start with chapter 6. As already seen, this chapter contains a decisive move in the argument that leads to Aristotle's theory of principles:

after beginning the positive part of the inquiry from an opinion shared by all his predecessors (chapter 5), in chapter 6 Aristotle accentuates the distance between the theories of principles advanced by his predecessors and the one he is progressively constructing. First of all, both the notion of *ousia* and the distinction between the various categories are used to move the investigation to the level of *ousia*, understood as the most general kind to which natural things belong. Moreover, these same tools are used to introduce the notion of substrate, by distinguishing it from the opposites. In particular, Aristotle employs the distinction between substance and attributes, the idea that attributes depend upon substance, and the idea that substance has no opposite to move from a generic functional distinction between opposites and substrate to the theses that opposites and substrate have a different being, that the substrate has priority over the opposites and that it is a principle (189a20–34).[79]

In chapter 7, where the search for the principles of natural things comes to an end, Aristotle also employs several theses that are tackled in the *Categories*: the distinction between substrate (*hypokeimenon*) and accidents, and the idea that only substance is not said of something else, whereas all other things are said of substance (190a33–b1). The use of these theses in chapter 7 is particularly relevant since it makes clear both the debt of *Physics* I to the *Categories* and its distance from it. In chapter 7, the theses just mentioned are employed at an important juncture of the inquiry, namely the passage from the arguments that show the presence of a substrate in accidental change to the arguments that extend the notion of substrate to substantial change.[80] The most evident function served by the theses just mentioned is to corroborate a point established in the preceding part of the chapter (189b32–190a31): in the case of accidental change, there is a substrate (*hypokeimenon*), understood as what comes to be (*gignomenon*) and as the starting point of the process, which (or part of which)[81] remains during the process and acquires a property that it lacked at the beginning of it. Broadly, the argument put forward at 190a33–b1 goes as follows: if the result of a process is an accidental property (e.g. a quality or a quantity), since accidental properties are said of a substance

[79] See section II.

[80] A different interpretation of this passage is advanced by Charles in this volume.

[81] It is controversial whether, in *Phys.* I.7, the physical substrate is conceived of only as a simple entity or also as a compound one. The simple-entity interpretation seems more consistent both with the *Categories*' notion of substrate and with the physical notion of matter, which eventually appears in *Phys.* I. However, the compound-entity interpretation is suggested by several passages of the book (e.g. 190a14–16, 190b23–7, 191b35–192a2).

(i.e. a substrate in the sense of a subject of predication), then – for the reasons given in the preceding part of the chapter (189b32–190a31) – it follows that, in all cases of this kind, there is a substance/substrate, which is the starting point of the change, which (or part of which) remains during the change, and which (or part of which), at the end of the change, possesses an accidental property that it lacked at the beginning of the change.

But how are things when the change at issue is the coming to be of a substance? In this case, the categorial tools are not helpful. Rather, they even seem to provide an obstacle. For, in so far as substantial generation is concerned, and in so far as one sticks to these categorial tools, the result of the process cannot be articulated into parts (into substrate and opposite/form), but is, rather, indivisible.[82] Perhaps, then, the theses from the *Categories* are mentioned also to emphasise this problem: they are mentioned to show that, in order to identify, within substance and substantial generation, a distinction between substrate and opposite, further tools (besides the categorial ones already employed both in chapters 6 and 7) are needed. Indeed, this is what Aristotle seems to aim at in the following passage:[83] he introduces new means and new points of view to show that the *result* of any change (both accidental and substantial) is something composite, which comprises an acquired property or an acquired

[82] It is worth noting that the conclusion drawn at 190b10–13 concerns not only the *terminus a quo* of change (190b12–17), but also the *result* of change (190b10–12), just like the initial categorial remark (190a34–b1). The conclusion drawn at this point (190b10–13) seems to be the final conclusion of what has been said from the beginning of the chapter, and hence also the final conclusion of the partial argument that starts at 190a31. On this reading, the argument is aimed at showing both that substantial change proceeds from a substrate and something that is opposed to the result (190b1–3, 190b9–10) and at showing that the result of substantial change is something composite (190a33–b1, 190b10–17). This second aim is introduced by the categorial remark at 190a34–b1. Arguably, its function is to open the route towards the formulation of Aristotle's theory of principles at 190b17–20, which is about the principles of substances viewed as *results* of change. On this reading, Aristotle arrives at the final conclusion that also the result of substantial change has a composite nature (i.e. it comprises a *hypokeimenon* and a *morphe*) by means of a two-step argument: (1) at 190b3–5 he points out that also substances come to be from something, conceived of as a starting point and as a subject of change; (2) at 190b5–9 he points out that the kind of changes that the starting points/subjects of substantial change undergo are similar, in some relevant respects, to the kind of changes that the starting points/subjects of accidental change undergo (i.e. the change is either a kind of alteration or a kind of increase or a kind of decrease, etc.). From this it follows that also substances are composite entities, and that also substances come to be both from a substrate and from something that is opposed to the result of the coming to be (190b12–13).

[83] On the implications of this passage for an assessment of the basis used by Aristotle for his positive claims about principles in ch. 7, see Charles in this volume.

substantial form, and something that acquires it, which in turn can refer either to a (primary) substance or to a substrate in the sense of matter (190b10–13).

In sum, these passages from chapters 6 and 7 show that the categorial distinction between substance and accidents (and the other related theses previously mentioned) are used in *Physics* I as fundamental tools to arrive at the distinction between substrate, in the sense of matter, and form. This heuristic relation is apparent also if viewed from the perspective of the *Categories*, i.e. if we consider the way in which the distinction between substance and accidents is presented in the *Categories*. In *Cat.* 4a10ff. Aristotle presents as the most proper characteristic of substance its being able to receive opposites, while remaining numerically one and the same. Moreover, all the subsequent treatment of this characteristic is focused on change and on the role of substance as permanent substrate of opposites within (accidental) change (4a29–34). Lastly, even if the *Categories* does not tackle principles and does not label substance as a principle,[84] it nevertheless seems to come close to conceiving it as such when it claims that primary substance is that without which nothing else would be (2b3–6c, 15–17). All this shows that the substance of the *Categories* is not only viewed as a subject of predication, but also as a *hypokeimenon* of change and as a principle (although, of course, it is not a *hypokeimenon* and a principle in the same sense developed in *Phys.* I.7).

Let us move on. What is the epistemic role, in *Physics* I, of the categorial theses mentioned so far? The answer to this question depends upon the meaning and status that we attach to them.[85] We may consider these theses as linguistic or as ontological assumptions or both, and maintain that the treatises included in the *Organon* are meant to provide either a logical or an ontological system (e.g. a systematic stage-by-stage exposition of logic or a kind of proto or early metaphysics), which is to be used in the subsequent scientific investigations: if so, we should conclude that *Physics* I relies heavily on doctrinal assumptions and on distinctively Aristotelian precast tools. Alternatively, we may attribute a less demanding status to the categorial theses used in *Physics* I: we may conceive them as dialectical principles, and maintain that *Physics* I only presupposes a familiarity with certain dialectical practices and rules, like those described in the *Topics* and in the *Sophistical Refutations*. On this view, the categorial theses used in *Physics* I should be considered as part of the dialectical practices used to

[84] However, see *Metaph.* Γ 2.1003b6.

[85] I am grateful to Alan Code for pressing me on this point.

distinguish the various meanings of words or the various kinds of things that a word may refer to.[86] So, from this point of view, dialectical practices and rules would be used in *Physics* I not only as a means to discuss and test proposed theories of principles, but also as a positive and constitutive basis (although not the sole one: cf. 190a31–b17 and relative comments above) to define the principles themselves.[87]

This is a big issue that goes beyond the scope of this essay.[88] Here I shall limit myself to two remarks, aimed at providing evidence respectively against the first hypothesis and in favour of the second hypothesis.

First, the hypothesis that the categorial distinctions and theses employed in *Physics* I are not conceived by Aristotle as parts of a logical and/or ontological system is suggested by the fact that, in spite of the clear use in *Physics* I of terms and concepts also employed in the *Organon*, there are also some noticeable differences, both terminological and doctrinal. The most evident cases include the meaning of the term *steresis* (privation), the distinction between *katholou* (universal) and *kath'ekaston* (particular), and the meaning of the expression *kat' hypokeimenou legetai tes ousias*.[89]

Second, the hypothesis that the categorial distinctions and theses employed in *Physics* I are conceived of by Aristotle as dialectical principles is suggested by the fact that the kind of resolution of Parmenides' argument for the unity of being, which Aristotle accomplishes in chapter 3, and which is carried out by means of the distinction between various senses or kinds of 'being', is presented in *SE* 33 as an application of dialectical principles.

We can now conclude this inquiry into *Physics* I's assumptions and relation with the *Organon* as follows. In so far as scientific knowledge,

[86] Dialectical – especially definitorial – practices are also at issue in ch. 6, in the argument for the reduction of opposites to one single primary pair (see Code in this volume).

[87] This interpretation is close to the one proposed by Wieland (1992).

[88] Both the two interpretations (which I mentioned above) of the function played by the *Organon* within Aristotle's writings have their supporters. In particular, the second one would fit the (apparently old) hypothesis that the *Categories* are 'what comes before the *Topics*' and are, together with *On Interpretation*, ancillary to the *Topics* (see Husik 1952: 96–112; Menn 1995; Burnyeat 2001: 109–10).

[89] *Steresis*: the notion of *steresis* employed in *Phys.* I is different from the one employed in the *Categories* (11b17ff., esp. 12a29–31); more radically, they are incompatible. *Katholou/kath'ekaston*: although the description of the method for discovering the principles provided in *Phys.* I.1 has some noticeable features in common with *APo* 71b29–72a5, the distinction between *katholou* and *kath'ekaston* employed in these two passages is antithetical: in *Phys.* I.1 the things more knowable to us and the things more knowable by nature are respectively presented as the *katholou* and the *kath'ekasta*, whereas in *APo* they are the other way round. *Kat' hypokeimenou legetai tes ousias*: this expression is used in *Phys.* I (185a32) in a context where the 'doctrine' of the categories is clearly at issue, but its meaning is different from the one employed in the *Categories* (1b10–12).

principles and method are concerned, *Physics* I relies on assumptions and views that Aristotle considers to be common to and shared by his predecessors. Further, besides being dialectically justified, most of these assumptions receive in *Physics* I a further – scientific – justification. Moreover, in so far as the categorial distinctions and theses are concerned, *Physics* I relies on principles and data that are tackled in the *Organon*, in particular in the *Categories* and in the *Topics*. If we consider these distinctions and theses as dialectical principles, rather than as part of a logical and/or ontological system, we may conclude that *Physics* I does not presuppose a doctrinal system but more simply a certain familiarity with dialectical practices and rules.

IV. Concluding Remarks: Looking Forward

In the foregoing, I have suggested that, although *Physics* I's inquiry is of a scientific and physical kind, it plays an introductory role within Aristotle's physical project, aimed at, as it were, setting its scene. This introductory role and its specific nature manifest themselves in various aspects of the book: the way in which it describes the central object of natural science (i.e. natural substances), the evidence it uses, its target, its heuristics, its peculiar relation with the tradition of natural philosophy, the way in which it identifies and conceives of the principles of natural substances (i.e. as the essences of principles as such), etc. Further, to clarify the role of *Physics* I and the kind of introduction it provides to Aristotle's physical writings, I explored the relation between *Physics* I and other parts of the corpus. In section III, I considered *Physics* I's background and focused on its relation with the *Organon*. In sections I and especially II, I addressed – among other things – the relation between *Physics* I and what comes next in Aristotle's physical enterprise. There I focused especially on the grounding role that the theory of principles argued for in *Physics* I has for the subsequent physical inquiries, both in the sense that it provides the foundation on which to build and in the sense that it is an outline to be filled in.[90] At various points I also addressed the issues about the role of *Physics* I's *argument* within Aristotle's broader physical project and about the epistemic status of the account of principles provided in *Physics* I. I emphasised some reasons to think that *Physics* I has epistemic priority over other Aristotelian texts and arguments on the same topic. However, it remains to be seen whether *Physics* I's argument actually succeeds in

[90] On this aspect of Aristotle's mode of inquiry see Quarantotto (2017: 19–42, 247–59).

providing the epistemic foundation for Aristotle's subsequent inquiries into principles, that is, whether it is sufficient to establish its conclusions (i.e. its account of principles) or whether, instead, it includes argumentative gaps. Further, it remains to be seen whether these supposed argumentative gaps are filled in by Aristotle in subsequent inquiries or not. Of course, the assessment of this issue depends greatly on how we interpret the theory of principles argued for in *Physics* I, that is, on the theoretical content and commitment we attribute to it. The more controversial this is, the more difficult it is to resolve the matter. That said, the issue is important enough not to leave it aside. Perhaps this is a good question for us to tackle in a future discussion, as Aristotle would say 'beginning from another starting point' (192b4).

Physics I.1

Andrea Falcon

[T1: 184a10–16] Since knowledge and science come about, in all the investigations in which there are principles or causes or elements, from grasping these things (for it is then that we think we know each thing, whenever we have grasped the first causes and the first principles and <got> as far as the elements), it is clear that in the science of nature too one should try to determine, first of all, the things concerning the principles. [T2: 184a16–21] The natural road is from what is more knowable and clearer to us to what is clearer and more knowable by nature: for it is not the case that the same things are knowable[1] to us and also <knowable> without qualification. For this very reason, it is necessary to proceed in this way from what is unclearer by nature but clearer to us to what is clearer by nature and more knowable. [T3: 184a21–3] What is initially evident and clear to us are the things that are confounded to a degree: it is only later, starting from these [confounded] things, that the elements and the principles come to be known to those who divide them. [T4: 184a23–6] For this reason, one must advance from the universals to[2] the particulars: for the whole is more knowable in relation to perception, and the universal is a sort of whole, for the universal comprises many things as[3] its parts. [T5: 184a26–b5] The same thing happens in a way to names in relation to their definition: <a name> signifies some whole in an indefinite way, e.g. 'circle', whereas its definition divides it into the particulars. Children too at first call[4] all men fathers and all women mothers, and later distinguish each of the two.

I. Three Initial Questions on *Physics* I

I have divided *Physics* I.1 into five short texts [TT1–5]. In the pages to follow, I will try to answer three questions. There are, needless to say, many

[1] *OCT*: γνώριμα. E: γνωριμώτερα. [2] *OCT*: ἐπὶ. E: εἰς. [3] *OCT*: ὡς. E: ὥσπερ.
[4] *OCT*: προσαγορεύει. E, V, I: ὑπολαμβάνει.

other interesting questions arising from this exceedingly difficult chapter, but the following three seem to be central to any reading of *Physics* I.1:

1. How are the objects of study, the principles of the science of nature, introduced in [T1]?
2. What is the precise nature of the road to the principles outlined in [TT2–4]?
3. How do the two examples in [T5] contribute to illustrating the progression from the universal (καθόλου) to the particulars (καθ' ἕκαστα) advanced in [T4]?

A few words on the exegetical strategy I will adopt in the pages to follow are in order. I will approach the text without presupposing any particular view on how *Physics* I.1 may (or may not) relate to other Aristotelian works. More directly, Aristotle begins the chapter by stating: "In the science of nature too one should try to determine, first of all, the things concerning the principles." The Greek word πρῶτον, which I have rendered as "first of all," is important. Evidently, Aristotle sees *Physics* I as the first, necessary step in a larger project. He calls this larger project "science of nature" (ἐπιστήμη περὶ φύσεως). In the pages to follow, I will not discuss how Aristotle conceives of the science of nature, or how our book as a whole may contribute to this science. Nor will I discuss how *Physics* I fits with the rest of the *Physics*.[5] I will also bracket a further question concerning the relation between our chapter and *APo* II.19, where Aristotle sketches out how we come to know the principles of a science.[6] To the extent that it is possible, I will try to develop a reading of *Phys.* I.1 solely on the basis of the conceptual resources introduced in *Phys.* I.1.

II. The First Question: The Principles

In [T1], Aristotle introduces his object of study as "the things concerning the principles" (τὰ περὶ τὰς ἀρχάς). The principles in question are those that pertain to the study of nature. Interestingly enough, Aristotle does not argue for the existence of principles of nature. Rather, *he is assuming their existence.* More precisely, he is assuming that the study of nature is an

[5] I refer the reader to Menn (forthcoming) for such a discussion.

[6] For an alternative exegetical strategy, see Bolton 1991. Bolton starts out his investigation by assuming that the aim of *Phys.* I.1 is the same as that of *APo* II.19, namely to explain how we come to know the principles of a science, and goes on to develop a unified account of the two chapters.

investigation (μέθοδος) which possesses principles. This assumption is sufficient to motivate us to search for the principles of nature because, as Aristotle himself explains, when we have grasped the (first) principles, we think we have knowledge.

I will return to the distinction between *principles* and *first principles* shortly. For the time being, I am content to note that in *Phys.* I.1 we are not told much about the principles beyond the fact that they are *principles of knowledge.* Aristotle's strategy may be usefully compared with Theophrastus'. The latter was reportedly dissatisfied with how Aristotle approached the question of the principles of nature in [T1]. In particular, Theophrastus felt that an argument was required for the claim that there are in fact principles of nature. In his own *Physics,* he tried to argue for the existence of principles of nature, starting from the objects studied by the science of nature. Philoponus and Simplicius are our sources of information for Theophrastus' argument:

> Natural things, [Theophrastus] says, are either bodies or have their existence in bodies (for instance, tendencies, capacities, and the like); but all bodies and things that have their existence in bodies are composite ... If, therefore, natural things are bodies or have their existence in a body, and these things are composite, then natural things are composite. But all composite things have elements, causes, and principles: for simple things are the elements of composite things. Hence, natural things have principles, causes, and elements. (Philop. *In Phys.* 4.12–15 and 4.34–5.3. Trans. Osborne, with modifications)
>
> But that there are principles of natural things, the whole subsequent argument will show, and it does not need to be demonstrated now; and it was for this reason I think that [Aristotle] passed over this premise. But Theophrastus, at the beginning of his own *Physics*, gave a proof of this too, saying: but that there are principles of natural things is clear from the fact that natural bodies are composite, and that everything that is composite has as principles the things of which it is composed. For everything that is natural either is a body or at any rate has body; and both of these are composite. (Simpl. *In Phys.* 9.5–10. Trans. Sharples)

This is not the place for a full discussion of the strategy adopted by Theophrastus.[7] What matters is that Theophrastus felt that an argument was needed for the claim that the science of nature has principles, and that he supplied an argument based on what he took to be the object of the

[7] The reader will find a full discussion of both passages in Laks 1998: 142–69, and de Haas 1998. Cf. Sharples 1998: 41–6.

science of nature, namely natural bodies and properties of natural bodies.[8] His argument goes as follows:

1. The science of nature is concerned with natural bodies and properties of natural bodies.
2. Natural bodies and properties of natural bodies are composite things.
3. All composite things have principles.
4. Natural bodies and properties of natural bodies have principles.
5. The science of nature is concerned with the principles of natural bodies and properties of natural bodies.

The strategy adopted by Theophrastus has at least one serious shortcoming: it presupposes that we have a precise idea of the object of the science of nature as we engage in a search for its principles. But it is emphatically not obvious that we have a grasp of the subject matter of the science of nature at this early stage of our investigation. Quite the opposite. The search for the principles is also a search for an adequate starting point in the study of nature. At the outset of *Phys.* I.2, Aristotle begins his search for the principles of nature with a review of the theoretical positions available on the topic of the principles, including the view that there is only one principle and that this principle is not subject to change. It does not take long to realize that Aristotle will have to dismiss this view before he is in a position to make any claim about the subject matter of the science of nature.

In [T1], Aristotle announces an investigation into the (first) principles of nature on the provisional assumption that there are such principles. This assumption is sufficient to motivate our investigation, because all knowledge, including knowledge about nature, depends on grasping the (first) principles. Aristotle begins his actual investigation at the outset of chapter 2 with the following words: "There must either be one principle or more than one." He concludes it only at the end of *Physics* I.9. He marks the end of his investigation as follows: "So, then, let us take it as determined in this manner that there are principles and which and how many in number they are." It is only at that point that a full answer to the question concerning the number and nature of principles is firmly in place. Moreover, it is only

[8] I note in passing that many, if not most, philosophers working within the Aristotelian tradition have adopted a similar description of the subject matter of natural philosophy on the basis of *DC* I.1 combined with *DC* III.1 (e.g. Avicenna, Averroes, and Zabarella). For more on this, I refer the reader to Falcon 2016.

at that point that the assumption made in the opening lines of *Phys.* I.1, *that there are principles*, is fully vindicated.

It is time for a closer look at how Aristotle proceeds in [T1]. This stretch of text begins with an ἐπειδή-clause. The subject is "knowledge and science" (τὸ εἰδέναι καὶ τὸ ἐπίστασθαι). τὸ εἰδέναι and τὸ ἐπίστασθαι need not be synonymous. In particular, ἐπίστασθαι can be a specific form of εἰδέναι. In other words, when Aristotle says "knowledge and science," we can understand "knowledge *in the sense of* science."[9] If we read the text in this way, then our text contains the following claim: knowledge in the sense of science comes about, in all the investigations (μέθοδοι) which possess principles or causes or elements, from knowing those things (sc. principles, causes, elements). Aristotle does not stop to elaborate on the investigations which have no principles, no causes, and no elements. But it is not difficult to find examples of such investigations. Star-gazing, which in the *Posterior Analytics* is presented as a fact-gathering inquiry not only distinct from but also subordinate to the science of astronomy, may be an investigation without principles, causes, or elements.[10] If taken in isolation from the subsequent explanatory stage, the nine books of the *History of Animals* may be considered another investigation without causes, principles, or elements. More directly, if we master the biological data collected in the *History of Animals*, we have εἰδέναι but not yet ἐπίστασθαι about animal life. For the latter, we need knowledge of the principles, elements, and causes – the sort of knowledge we acquire only at the explanatory stage of the study of animals.[11]

Aristotle is assuming that the study of nature is one of the investigations that have principles, causes, or elements. If we follow him in assuming that the study of nature is one of these investigations, then we are required to

[9] For a defense of this reading, which goes back at least to Alexander of Aphrodisias (as reported in Simpl. *In Phys.* 12.14–13.13), I refer the reader to Burnyeat 2011.

[10] *APo* I.13.78b38–9.

[11] It is worth pausing to note that the distinction between fact-gathering and explanation-giving investigations in my two examples may not be quite the same. The point Aristotle seems to make in *APo* I.13 is that one needs special skills and knowledge to gather facts about the stars – skills and knowledge that the mathematical astronomer may not have (and vice versa). But it is far from obvious that the skills that the animal observer (and compiler of the *History of Animals*) has are basically distinct from those of the animal explainer. Consider the most recent research on the *History of Animals*. This research has shown that the *History of Animals* does not consist in a simple collection of data; rather, the data are already *organized* for their subsequent explanation. Cf. Gotthelf 2012 (esp. ch. 12: "Data-Organization, Classification, and Kinds: The Place of the *History of Animals* in Aristotle's Biological Enterprise"). But as soon as a concern for explanation is built into the *History of Animals*, it becomes really difficult to claim that the skills and knowledge involved at the pre-explanatory stage of investigation are basically different from those that are required at the explanatory stage.

look for them. Why? Because, as Aristotle explains in the subsequent γάρ-clause, "it is then that we think we know (γινώσκειν) each thing, whenever we have grasped (γνωρίσωμεν) the first causes and the first principles and <got> as far as the elements." The γάρ-clause is open to more than one reading. For one thing, it is not clear whether Aristotle is stating his own particular view about knowledge, or whether he is presenting a more general conception of knowledge. It depends, ultimately, on how we understand "we think." I am not inclined to take it to be equivalent to "we [in our school] think," or "we [in this room] think," or even "I, Aristotle, think." First, Aristotle is not making explicit contact with his own views as stated elsewhere (most notably, in the *Posterior Analytics*). Second, in the rest of *Physics* I, when Aristotle wants to call attention to the fact that he is speaking in his own voice, he typically adds ἡμεῖς.[12] For another thing, Aristotle speaks of "first principles" and "first causes." These expressions can be taken to be equivalent to "proximate principles" and "proximate causes." But they can also be taken to mean "ultimate principles" and "ultimate causes." In the latter case, the thought would be that the person who has mastered any investigative domain with principles, causes, or elements is expected to go all the way up to the ultimate principles and the ultimate causes and as far as the elements.

Which of the two readings of "first principle" and "first cause" is to be preferred has an impact on how the search for the principles is to be understood. Is Aristotle announcing an investigation that is expected to result in the discovery of the ultimate principles? Or is he rather thinking of an inquiry into the principles whose goal is to get to the proximate principles? Consider how, in the rest of *Physics* I, Aristotle casts the search for the principles conducted by his predecessors. By his lights, they were never entirely clear on the distinction between principles and first principles. They adopted the language of opposition and agreed that the opposites are principles, but they failed to distinguish between opposites and primary opposites. Aristotle tells us that this distinction is very important. The primary opposites alone have the status of first principles because (a) they cannot be derived from one another, and (b) they cannot be derived from anything else, whereas (c) everything else can be derived

[12] See, for instance, the opening lines of *Phys.* I.7: "This then is what *we ourselves* say" (ὧδ' οὖν ἡμεῖς λέγωμεν) (189b30–1). Cf. also how Aristotle contrasts his views to those of others in *Phys.* I.8 and I.9: "*They* acquired this opinion for the reasons which have been stated. *We*, however, say that . . ." (ἐκεῖνοι μὲν . . . ἡμεῖς δὲ λέγομεν) (191a33–4); "For *we ourselves* say that matter and privation are different . . . *but these other people* <say that> the great and the small are alike what is not" (ἡμεῖς μὲν . . . φαμεν . . . οἱ δὲ) (192a3–8).

from them.[13] In *Physics* I, Aristotle places himself in continuity with the search for the principles conducted by his predecessors. His inquiry into the principles is meant to emerge from what was accomplished by his predecessors. At the same time, it goes emphatically beyond anything done by them. One way in which his investigation goes beyond is that it is conducted on a firm grasp of the distinction between principles and first principles. It is tempting to see this distinction already foreshadowed in the opening lines of *Physics* I. That is, when Aristotle refers to "first principles," he is not just referring to proximate principles. Rather, he is indicating an inquiry that will lead from proximate principles to deeper principles that cannot be derived from each other or from anything else – and principles from which can be derived everything else.

III. The Second Question: The Road to the Principles

If Aristotle adopts the approach suggested above, the transition from [T1] to [T2] is relatively smooth and straightforward. More directly, if we accept that proper knowledge comes from knowing the principles, the obvious question is: *How do the principles become known?* This is a general question, but it is an especially pressing question for us, given that we are about to engage in the study of nature, and proper knowledge in this field of study is contingent on grasping the relevant principles.

[TT2–4] contains the answer to the question how the principles come to be known. The strategy adopted in this stretch of text is one of a progressive illumination in three steps of a single epistemic journey. It is especially important that each step builds on the previous one. In particular:

[T2] gives us the *terminus a quo* and the *terminus ad quem* of our journey. While the former is described as "what is more knowable and clearer to us" (ἐκ τῶν γνωριμωτέρων ἡμῖν καὶ σαφεστέρων) the latter is identified with "what is clearer and more knowable by nature" (ἐπὶ τὰ σαφέστερα τῇ φύσει καὶ γνωριμώτερα).

[T3] introduces the mechanism that allows us to move from one class of things to the other: it is by means of division (διαίρεσις) that we come to the things that are more knowable and clearer by nature. We will have to understand how Aristotle conceives of division. For the time being, suffice it to note that "the things that are confounded to a degree" (συγκεχυμένα μᾶλλον) are identified with "what is initially evident and clear to us."

[13] I am relying on what Aristotle says on the pre-Socratic use of the opposites as principles in *Phys.* I.5.

Finally, [T4] casts the entire movement as a process from the καθόλου to the καθ' ἕκαστα (for the time being, I leave these words untranslated).

At least two observations are in order. The first observation is this: we are not simply expected, or advised, to start from the things that are more knowable to us; rather, *we are required to do so*. We have no explicit argument for this requirement. Consider, however, the emphatic position of πέφυκε at the beginning of [T2]. In all probability, this position is designed to call special attention to the fact that the road that goes from what is more knowable to us to what is more knowable by nature is a natural one. Since it is natural, Aristotle may think that this road is open to all of us.[14] And yet Aristotle may also think that there is no alternative to it. He may think, in particular, that this road is forced upon us by our human nature.[15] Which would explain why he says that "it is necessary to proceed in this way," namely "from what is unclearer by nature but clearer to us to what is clearer by nature and more knowable."

The second observation is that Aristotle describes the epistemic journey in terms of clarity: we move from what is clearer to us to what is clearer by nature. This case is not unique in the Aristotelian corpus. I would like to draw attention to two other passages where the link between knowledge and clarity is established. The first comes from the treatise *On the Soul*. At the outset of *On the Soul* II.2, we are told that "it is from things that are unclear but more evident [to us] that what is clear and more knowable in relation to reason comes to be" (413a11–12: ἐκ τῶν ἀσαφῶν μὲν φανερώτερων δὲ γίνεται τὸ σαφὲς καὶ τὸ κατὰ λόγον γνωριμώτερον). The second is a key methodological passage controlling the overall argument advanced in the *Eudemian Ethics*. In *Eudemian Ethics* I.6, Aristotle is about to embark on an investigation of key concepts such as happiness and virtue. He describes his investigation as a progressive clarification of what is said truly but not clearly (1216b32–3: τῶν ἀληθῶς μὲν λεγομένων οὐ σαφῶς δέ). The end result is secured by constantly replacing what is usually said in a confounded way with what is more knowable (1216b33–5: μεταλαμβάνουσιν ἀεὶ τὰ γνωριμώτερα τῶν εἰωθότων λέγεσθαι συγκεχυμένως). In this case, Aristotle is forthcoming on the

[14] Cf. *Metaph.* Z 3.1029b3–5: "Learning happens *for all* in this way: from things that are less knowable by nature to things that are more knowable." But this does not mean, I hasten to add, that Aristotle envisions this road as a *reliable* route to the principles. On the contrary, the review of what Aristotle's predecessors accomplished, or rather failed to accomplish, suggests that the road to the principles is a long and arduous road to take, and one in which a great deal can go wrong.

[15] In this sense, this is also the *right* or *proper* route to the principles, namely the route that we *ought to take* if we want to get to the principles.

main reason for adopting this procedure: "each person has some affinity with the truth" (1216b30–1: ἕκαστος οἰκεῖόν τι πρὸς τὴν ἀλήθειαν).[16]

It is time to look a bit more closely at how the epistemic journey is described in *Phys.* I.1. First of all, I would like to explore the starting point of this epistemic journey. According to Aristotle, the road to the principles begins from "the things that are confounded to a degree" (τὰ συγκεχυμένα μᾶλλον). The key word is συγκεχυμένα, which I have rendered with "confounded." A passage from *HA* I.15 helps us shed some light on what Aristotle has in mind. There, Aristotle is recalling his well-known thesis that living bodies display up to three functional dimensions: up/down, front/back, and left/right. These functional dimensions are most clearly articulated in the human body because of its erect posture. The human body is the exception rather than the rule. The other animals either do not have all three dimensions or "they do have them, but have them confounded to a degree" (τὰ δ' ἔχει μὲν συγκεχυμένα δ'ἔχει μᾶλλον).[17] Aristotle does not mean to say that these functional dimensions are not present in the second group of animals; rather, he means to say that they are found in the same place. However, we are able to distinguish between the front and the top of a living body because we can trace these functional parts to different capacities of the living body. For Aristotle, the front is where the sense-organs are implanted, whereas the top is the entry point of nourishment. This distinction may not be evident to sense-perception, but it is always *clear to reason*. Put differently, and more boldly, we have a *biological theory* that enables us to bring to light the functional organization of the living body.

The passage from *History of Animals* I.15 suggests that the epistemic journey outlined in *Phys.* I.1 need not be cast in psychological terms as the progressive clarification of what is initially confused. In other words, it is not necessary to think that the starting point of our epistemic journey is a psychological state of confusion. We are not confused about the functional dimensions of a living body. Rather, the functional dimensions are found in the same place in a living body. Hence, they are confounded. It is this

[16] I am adopting the translation printed in Inwood and Wolf 2013. Simpson 2013 has a good alternative: "each person has some familiarity with the truth." Kenny 2011 renders our passage as follows: "everyone has something of their own to contribute to the truth." This rendering does not seem to be right because it makes the truth an intersubjective rather than an objective fact. Again, from the fact that we have some affinity or familiarity with the truth, it does not follow that attaining the truth is a foregone conclusion (see note 14).

[17] *HA* I.15.494a32. I note, in passing, that this passage is textual evidence for taking μᾶλλον together with συγκεχυμένα.

state of *confoundedness* – a state in which different things are found in the same place – that Aristotle envisions as the starting point of the epistemic journey that leads us to the principles.[18]

If one takes this idea seriously, then the epistemic journey that Aristotle outlines in *Phys.* I.1 is a rational process that entails the progressive articulation of what is initially confounded to a degree. What is initially confounded reveals its complexity and structure as we analyze it. Since Aristotle speaks of division (διαίρεσις), we can safely assume that our analysis consists in drawing distinctions and separating things that are initially found together. Our analysis reaches its natural end when we have reached the maximally clear. Relative to what we are trying to understand, the latter does not admit of any further articulation or distinction.[19]

In the Aristotelian tradition, [TT2–3] is sometimes kept separate from [T4]. On this reading, after having sketched out the method by which we *learn* the principles, Aristotle would present the method by which we *teach* them. The καθόλου introduced in [T4] would be the sort of thing that we reach at the end of the process outlined in [TT2–3]. This καθόλου would be something that is *clearer and more knowlable by nature*. Moreover, it would be *the last in the order of discovery* and *the first in the order of teaching*.[20] It is not difficult to find an exegetical motivation for

[18] A passage from *Generation of Animals*, registering a putative phenomenon relevant to the transmission of certain physical traits from the parents to the offspring, confirms the non-psychological reading of τὰ συγκεχυμένα: "in some cases when the parents have wound-scars some of the offspring have gotten an outline of the scar in the same places, and in Chalcedon a tattoo on the arm of the father was marked on the child, though the figure was confounded and not clearly articulated (συγκεχυμένον καὶ οὐ διηρθρωμένον τὸ γράμμα)" (*GA* I.17.721b30–6).

[19] The reader of *Phys.* I.1 will find that both the psychological and the non-psychological readings are equally well documented in the scholarly tradition. Without the ambition of being exhaustive, I report hereafter a few translations of [T3], with emphasis on their rendering of τὰ συγκεχυμένα: Charlton 1970: "The things which are in the first instance clear and plain to us are rather those which are *compounded*. It is only later, through an analysis of these that we come to know elements and principles"; Bolton 1991: "The things that are first evident and clear to us are *things that are rather jumbled up*. Later on from them their elements become intelligible and the principles serve to separate those [jumbled up] things"; Hardie and Gaye 1930: "What is to us plain and obvious at first is rather *confused masses*, the element and principles of which become known to us later by analysis"; Ross 1936: "The path leads from what is better known to us to what is better known by nature. This path we must now follow: we must reach the first principles by analysis of *the confused data* we start with"; Pellegrin 2002: "ce qui est d'abord evident et clair pour nous ce son plutôt *les* <*ensembles*> *confus*; mais ensuite à partir de ceux-ci, deviennent connus, pour qui les divise, leur éléments et leur principes"; Stevens 2012: "Or les choses qui nous sont d'abord manifestes et claires sont *les plus confuses*, mais plus tard, à partir de celles-ci, les éléments nous deviennent connus, pour qui les divisent."

[20] This reading is adopted by Averroes. But it is emphatically not confined to Averroes and the tradition that goes back to his long commentary on the *Physics*. An introduction to this reading and its fortunes in the Arabic tradition can be found in Lettinck 1994: 91–109. For a recent treatment of the reception of our passage in the Arabic and Latin tradition, see Cerami 2009.

this reading. In the Aristotelian corpus, the expressions καθόλου and καθ' ἕκαστα are normally used to capture the contrast between universal and particular. This is the contrast that exists between genus (animal) and species (human being), and between species (human being) and individuals (Socrates, Callias). By inserting a break after [TT2–3], and taking [T4] to introduce a new question, namely the question of how we *teach* the principles we have just discovered, it is possible to assume that the usual contrast between general and specific is at work also in *Phys.* I.1.

I do not accept this reading. I do not see any textual evidence in support of the view that Aristotle has moved away from the question of how we *learn* the principles of the science of nature. As I have already suggested, I take the whole stretch of text in [TT2–4] to be concerned with the process by which we come to know these principles. By my lights, this is a single process that is illuminated in three consecutive steps. [T4] contains the third and final step. On my reading, [T4] builds on [TT2–3],[21] and the καθόλου is something that is *clear and evident to us but still unanalyzed.* It is by means of division that we proceed from this καθόλου to the καθ' ἕκαστα.

It is sometimes said that Aristotle employs καθόλου and καθ' ἕκαστα in a peculiar or non-standard way in *Phys.* I.1. Of course, the implicit assumption is that there is a standard use of these terms, and that this use is codified at the outset of *On Interpretation* 7, where Aristotle contrasts καθόλου and καθ' ἕκαστον and calls universal (καθόλου) that which is predicated of many things.[22] I do not deny that καθόλου and καθ' ἕκαστα are terms of art in Aristotle's logic, but I dispute that Aristotle is assuming or even using his logical notions at the outset of *Physics* I. What we have seen so far suggests the opposite. For instance, we have seen that division (διαίρεσις) is not used in a technical sense to refer to the Academic method of division, let alone to a specific version of that method. Rather, it is used in a more general sense to indicate that, in the course of our inquiry, we will have to draw distinctions or to separate things that are initially confounded. We have also seen that it is far from obvious that Aristotle

[21] The use of the conjunction διό is noteworthy: it suggests that Aristotle is building on what he has just said.

[22] *Int.* 7.17a39–b1: "I call universal (καθόλου) that which is by its nature predicated of many, and particular (καθ' ἕκαστον) that which is not; for instance, human being is among the universals and Callias among the particulars." The reader should keep in mind that Aristotle uses καθ' ἕκαστον to refer not only to individuals such as Socrates and Callias but also to the so-called *infimae species* (species under which there are only individuals). For this latter use see, e.g., *APr* II.23.68b20 and *APo* II.13.97b28.

is trying to make contact with his own conception of scientific knowledge as outlined in the *Posterior Analytics*. Rather, he may be invoking a conception of knowledge that is acceptable well beyond the narrow boundaries of his school. After all, the claim that knowledge depends on having grasped the first causes *or* the first principles *or* the elements is not specific enough. In sum, the overall impression is that, in outlining the task that lies ahead of us in *Physics* I, Aristotle is casting his net as wide as possible.

With the above considerations in place, we can look closer at what Aristotle says in [T4]. In this stretch of text, Aristotle offers an argument for the conclusion that we must proceed from the καθόλου to the many things it comprises (the καθ' ἕκαστα). Here is a possible reconstruction of his argument:

1. The natural road is the one that goes from the things that are more knowable to us to the things that are more knowable by nature.
2. The things that are more knowable to us are more knowable to sense-perception.
3. The whole is more knowable to sense-perception than its parts.
4. The καθόλου is a kind of whole.
5. Therefore, we must advance from the καθόλου to the καθ' ἕκαστα.

The identification of the things that are *more knowable to us* with the things that are *more knowable to sense-perception* remains implicit in [TT2–3]. I have made it explicit in premise 2.[23] Premises 3 and 4 are derived from [T4]. How premise 4 is to be understood is not completely clear. In particular, the τι in line 184a25 and the ὡς (Ms E: ὥσπερ) in line 184a26 are ambiguous. Aristotle may be saying that (a) the καθόλου is a whole and the many things it comprises are its parts, or that (b) the καθόλου is *analogous* to a whole and the many things it comprises are *analogous* to parts. The second reading is sufficient to generate an argument to the conclusion that we must proceed from the καθόλου to the καθ' ἕκαστα. I recommend preserving the ambiguity of the Greek text. Aristotle may be trying to speak to different conceptions of the scientific inquiry. While some of his pre-Socratic predecessors conceived of the

[23] This identification is explicitly made in a number of other passages. Cf., for instance, *APo* I.2.72a1–5, where Aristotle identifies what *is more knowable to us* with what is *closer to sense-perception*: "Things are prior and more familiar in two ways: for it is not the same to be prior by nature and prior in relation to us, nor to be more familiar and more familiar to us. I call prior and more familiar in relation the things that are closer to sense-perception, prior and more familiar *tout court* the things that are further away" (tr. Barnes slightly modified). Cf. also *Top.* VI.4.141b5–14.

scientific inquiry as an analysis of concrete wholes into their material constituents, Plato and the Academic philosophers had a more abstract understanding of the search for the first principles. For them, the latter were not material principles such as earth, water, air, and fire. Aristotle may have crafted this passage in order to appeal both to those who understand their task as that of analyzing concrete wholes into their material constituents and to those who have a more abstract conception of their task.[24]

On the proposed reading, the καθόλου is the starting point of our investigation. Because it is (*analogous* to) a whole and, furthermore, because it comprises many things that are (*analogous* to) its parts, we are required to articulate its internal complexity. It is a complexity that is not clear to us. If the καθόλου is given to us by perception, then the καθ' ἕκαστα are things that are not evident at first. They become clear at a subsequent stage *by means of rational analysis*. Premise 3 helps us establish this point: since the καθόλου is (*analogous* to) a whole, it is more knowable to perception than the many things it comprises.

The following illustration might capture what Aristotle has in mind. Let us suppose that we hear someone speak, and hear each of the spoken words clearly. Each spoken word is a καθόλου that is *clear and evident to us*. It is clear and evident to us because it is *clear and evident to sense-perception*. Yet this does not mean that we can separate each spoken word into its syllables and, in turn, each syllable into its constitutive elements, namely vowels and consonants. Quite the opposite. Each spoken word we hear is like a whole, and as such it is at first indeterminate and confounded to a degree. The natural road is to proceed from this indeterminate and confounded καθόλου to its constitutive elements. The latter are the καθ' ἕκαστα. While the καθόλου is immediately available to perception, the καθ' ἕκαστα are obtained through analysis and by means of reason (λόγος). This is the sort of analysis and λόγος that is exemplified by someone who has mastered the art of grammar. This person is able to move from the spoken words to their constitutive elements and back. Of course, this person is also able to give an account of the intermediate step(s). As this example shows, we cannot proceed to the constitutive elements unless we are able to separate the spoken words into their syllables.[25]

[24] This is of course in line with the overall impression that Aristotle is casting his net as wide as possible.

[25] I modeled my example after what Plato says on the discovery of writing in *Phil.* 15b–18d. In Plato's story, the starting point is spoken sound (φωνή). The articulation of spoken sound requires the discovery of the στοιχεῖα. The στοιχεῖα are our ἀρχαί, and the art of grammar (γραμματικὴ τέχνη)

In my illustration, the movement is double: from the spoken words to the στοιχεῖα and back. This point can be generalized with the help of a passage from the *Nicomachean Ethics*:

> We must notice the difference between arguments from principles and arguments to principles. For indeed Plato was right to puzzle about this when he used to ask if the road (ὁδός) is from principles or toward them – just as on a race-course (ἐν τῷ σταδίῳ) [we may proceed] from the starting line to the far end, or back again. For while we should certainly begin from things that are knowable, things are knowable in two ways: for some are knowable to us, while some are knowable without qualification. Presumably we should begin from what is knowable to us. (*NE* I.4.1095a30–b4)

Here too Aristotle is concerned with the movement from one class of things, things that are (more) knowable to us, to another class of things, things that are (more) knowable without qualification. However, there are some interesting differences between this passage and what Aristotle says in *Phys.* I.1. The most important and obvious one is this: there is no equivalent in *Phys.* I.1 to the image of the race-course employed in *NE* I.2. This image points to the fact that the movement is expected to be double – from what is (more) knowable to us to what is (more) knowable in itself, and back to what is (more) knowable to us. In other words, getting to the principles is not a goal in itself; rather, it is a station on a journey. If we reflect a moment on this point, we should find it uncontroversial: it is not that we have knowledge when we know the first principles but rather when we know the path *to* and *from* those principles. It is only at that point that we have mastered a field of study. We can press this point further, in my view. Consider, in particular, the following reformulation of our epistemic journey and its goal: "we start from what is better knowable to us *which we wish to know with a knowledge [that is] derived from the principles.*"[26] The emphasis on the second part of his sentence is mine. It draws attention to the fact that, on a possible reading of the double movement to the principles and back, the things that are more knowable to us remain the ultimate goal of our investigation throughout the entire

is the bond that unifies them all. A version of my example can be found in Themistius, who is clearly inspired by what Plato says in the *Philebus*. Themistius distinguishes between hearing a word and stating the elements from which the word is composed. The latter task requires a rational analysis of what is presented to sense-perception and is accomplished by a person who has mastered the art of grammar (in other words, a γραμματικός). Cf. Themist. *In Phys.* 1.7–11.

26 This is a quotation from Wieland 1975: 129. This article offers a summary of the main results reached in the first chapter of Wieland 1970. There is a great deal that I do not find congenial in Wieland's reading of *Physics* I, but this quotation certainly captures a salient feature of Aristotle's strategy.

process. It is on account of them that we engage in a search for the relevant principles. This thought can be reformulated by invoking the equivalence between things that are more knowable to us and things that are closer to sense-perception. Our journey begins with the things that are closer to sense-perception. It is because we want to know them that we engage in the search for the relevant principles. In this sense, the things closer to perception are not just the starting point of our inquiry; they are also the end of the entire scientific enterprise.[27]

IV. The Third Question: The Two Examples

In [T5] Aristotle gives us two examples to illustrate how we proceed from the καθόλου to the καθ' ἕκαστα. The first example goes as follows: "<a name> signifies some whole (ὅλον τι) in an indefinite way, e.g. 'circle', whereas its definition divides it into the particulars (καθ' ἕκαστα)." This example can be read in two ways.

The first reading starts from the observation that a number of different things can be named κύκλος in ancient Greek, including a geometrical κύκλος, an epic κύκλος, and the κύκλος of the zodiac. From this observation, this reading concludes that κύκλος signifies some whole (ὅλον τι), namely the collection of all the things that are named κύκλος. In particular, the name would signify in an indefinite manner (ἀδιορίστως) precisely because it would refer to different things that are lumped together by the name. It is only when we have the definition that we are able to analyze our initial whole into the relevant καθ' ἕκαστα.

There are at least two objections to this reading.

First objection: the things that are lumped together by the name κύκλος have only that name in common. If we were asked to define them, we would have to give one *logos* for geometrical κύκλος, another *logos* for epic κύκλος, and yet another *logos* for the κύκλος of the zodiac. But this would mean that the initial καθόλου would disappear as soon as we realized that the name κύκλος refers to vastly different things. This is a rather unexpected development. We have not been prepared for this possibility in [T4]. This first objection is not fatal, however. It is open to us to suggest that our example is concerned with the definition of the *name* κύκλος. When therefore Aristotle says ὁρισμὸς αὐτοῦ in 184b3, we should

[27] In *De caelo*, Aristotle is explicit in stating that the authoritative data available to perception are the end of the entire scientific enterprise: "end of productive knowledge is the product; *[end] of natural science is always what authoritatively appears to be so to sense-perception*" (III.7.306a16–17).

understand ὁρισμὸς τοῦ ὀνομάτος. I note, in passing, that this is how Ross seems to understand our example.[28]

Second objection: the καθόλου from which we are required to proceed is expected to be a whole that is clear and evident to sense-perception. This expectation can be derived from [T3]. But it is emphatically not obvious that the prima facie whole we obtain by collecting all the things that are named κύκλος is clear and evident to sense-perception. In fact, the opposite is the case. To sense-perception, a geometrical κύκλος, an epic κύκλος, and the κύκλος of the zodiac do not seem to constitute a genuine whole. We get a prima facie whole only if we remain at the level of language and pretend that sense-perception does not tell us that we use the name κύκλος to refer to vastly different things. This second objection can be rephrased as follows: it should be sense-perception that establishes the καθόλου and thereby gives us the starting point relative to us of the investigation into the principles. Of course, this does not mean that we are barred from starting from language, but a mere accident of language should not suffice to get us on the road to the principles.

With these two objections in place, it is time to turn to the second reading of our example. This reading takes its lead from the fact that the καθόλου is expected to be something clear and evident to us. Accordingly, it starts from the observation that the name κύκλος signifies some whole accessible to sense-perception: for instance, a circle drawn in the sand or on a blackboard. This name would signify the whole in an indefinite manner because the name alone does not give us access to the internal complexity of the circle. When we know how to use the name κύκλος, we are able to point to a circle drawn in the sand or on the board and say: "that is a κύκλος." However, that may be all we are able to say about the circle. On this second reading, we are able to analyze the circle into its elements when we are in possession of its *geometrical* definition.[29] More directly, it is only when we know this definition that we understand what sort of figure the circle is. Moreover, it is only then that we have access to the καθ' ἕκαστα of the circle. In our example, the καθ' ἕκαστα might be the constitutive elements of a circle, namely a point for the center and a line for the radius,

[28] Cf. Ross 1936: 458: "τὰ καθ' ἕκαστα seems to have here an unusual meaning: i.e. to mean the various senses of an ambiguous term. Though it is essentially the business of definition to state the logical elements of a complex term, incidentally in doing this it will distinguish the various meanings of the term, if this happens to be ambiguous."

[29] Euclid defines a geometrical circle as a "plane figure contained by one line such that all the straight lines falling upon it from one point among those lying within the figure are equal to one another" (*Elements* I def. 15).

or they could be the curved line that consists of all the points that are at the same distance from the point that we call center plus that point.[30]

Many details of the first example remain obscure, but the following general point is clear: the example is meant to illustrate the existence of two different epistemic stages, which are introduced by looking at their corresponding objects. While at stage one we grasp an undifferentiated καθόλου, at stage two we are able to define the καθόλου in terms of the relevant καθ' ἕκαστα. The first stage corresponds to a pre-technical or pre-scientific kind of cognition; the second corresponds to the cognitive state of someone who has fully mastered the science of geometry.

In a highly compressed text as this, it is very unlikely that Aristotle gives two examples to make a single point. The first example says nothing about how we move from one stage to the other. I submit that the second example may serve to fill this lacuna. In other words, while the first example introduces two cognitive stages in a single and continuous process of progressive articulation of what is initially confounded to a degree, the second example illustrates what may happen as we try to move from one stage to the other. I would like to elaborate on this idea by first making explicit the assumption that is implicit in the second example.

Aristotle thinks of the search for the principles of a science as a progressive articulation of what is initially confounded. He also conceives of this progressive articulation as a natural process. Recall, in particular, the emphatic position of πέφυκε at the beginning of [T2], which frames the entire process outlined in [TT2–4] as a natural one. Among other things, this means that the road to the principles as envisioned by Aristotle unfolds according to a common and natural pattern, which may mirror what happens at the early stages of our cognitive development. If this is correct, then we can use what happens at the early stages of our cognitive development as an illustration of what happens in the acquisition of the principles of a science. This seems to be the working assumption in the second example, which goes as follows: "Children too at first call all men fathers and all women mothers, and later distinguish each of the two."

In order to see what Aristotle may have wanted to illustrate with his second example, we need to recall that part of what we are expected to do in our search for the principles is to draw distinctions and to separate

30 A final note is in order. Whether we adopt the first or the second reading of the first example, we have to admit that καθ' ἕκαστα is not used to indicate the individuals instantiating the universal or the *infimae species* falling under the genus. In other words, the first example confirms that Aristotle is not using his logical apparatus in *Phys.* I.1.

things that are initially confounded. For instance, one thing that we are expected to do is to distinguish the role that the opposites play as principles of changeable things from the role played by the subject that undergoes change. According to Aristotle, this is a distinction that all his predecessors failed to make. In other words, they all failed to see that what changes is something that is initially confounded but at a closer look can be articulated into two things: the subject that undergoes change (man) and the opposite (uneducated). When a man becomes educated, the second element, which is always a kind of privation, is replaced by the other opposite (educated). At least in this respect, Aristotle's predecessors are like the children in our second example. The latter fail to distinguish between the role of the father and that of man, and between the role of the mother and that of woman. As a result of this, they call all men fathers and all women mothers.[31]

In the second example, Aristotle draws attention to an interesting mistake that children make at an early stage of their cognitive development. But what exactly is the mistake that they make? They certainly do not make a *perceptual* mistake, as Ross seems to think.[32] In other words, these children do not fail to discriminate their father and their mother from all other men and women. On the contrary, we must assume that the normal operations of their senses allow these children to tell their parents apart from one another and from all other women and men. Hence, the children who call all men fathers and all women mothers must make some other mistake. I would like to suggest that they make a *conceptual* mistake. The early thinkers too made this mistake as they were looking for the principles of natural things. In both cases, it is a mistake committed by reason. To see why, we need to recall that sense-perception is expected to get us on the road to the principles by providing a reliable starting point.

[31] Here I have gone beyond the narrow boundaries of *Phys.* I.1 in order to give a first, rough idea of how the procedure for the discovery of the principles outlined in our chapter may relate to the actual investigation carried out in the rest of the book. Of course, we have every reason to expect Aristotle to adopt this procedure in his subsequent discussion. But a full review of how we can use the first chapter to shed light on the argumentative strategy adopted in *Phys.* I goes emphatically beyond the limits of this chapter, so I am content to make one additional point. By my lights, what Aristotle starts doing in *Phys.* I.6 and is completed only at the end of *Phys.* I.7 can be described, with the help of the language introduced in [T4], as the articulation of a καθόλου into the relevant καθ' ἕκαστα. I diverge from Bolton 1991, esp. 26–7. Bolton thinks that the analysis of what is confounded to a degree starts only in *Phys.* I.7. His reading reflects the widespread assumption that *Phys.* I.7 is a more or less free-standing chapter. For a recent attempt to rethink the place of this chapter in the argument of *Phys.* I, see Kelsey 2008.

[32] Cf. Ross 1936: 457: "the child who recognizes the general appearance presented by all men and that presented by all women, without noticing the special appearance of its father and its mother."

Aristotle describes our starting point as things that are clear and evident to us but still confounded to a degree. It is from things of this sort that we get, by means of a rational analysis that entails disentanglement of what is initially confounded, to the relevant principles. It would be very surprising (to say the least) if Aristotle tried to undermine sense-perception by suggesting that children cannot *perceptually* discriminate their mother and their father from all the other women and men. What must be missing at the early stages of our cognitive development is the capacity to employ the relevant concepts.[33] As a result, they are not able to draw some important distinctions. In particular, they fail to see that one and the same person can be father and man, or mother and woman. Aristotle thinks that his predecessors committed similar mistakes as they engaged in a search for the principles.

One final comment is in order. In the ancient commentary tradition we find a *third* example, which is said to be equivalent to the *second* example given by Aristotle: it is like when we say that the thing that is approaching is an animal; when it comes nearer to us, we say that it is a human being, and finally that it is Socrates.[34] The second and the third example are emphatically not equivalent. In the third example, we have a web of concepts in place, and we know how to apply them. Due to a contingent situation (distance), we are not able to say whether the thing that is approaching is a human being or a dog. However, we do have the conceptual resources to say that the thing that is approaching is a human being and not a dog. In the second example, which is Aristotle's example, there is no such web of concepts in place. As children grow up (by learning a language), they eventually become able to perform what is envisioned in the third example.

[33] That Aristotle is concerned with how we (mis)apply concepts is made explicit in the textual tradition that reads ὑπολαμβάνει instead of προσαγορεύει.

[34] Philoponus has an extensive discussion of the third example, which he takes to be equivalent to Aristotle's example: "when we see Socrates from a distance, we know first that it is an animal, and in this respect the knowledge (γνῶσις) of him as an animal is indiscriminate (συγκεχυμένη): for an animal might be any one of the other animals. And then a human being works the same way; even if it is distinguished from other animals, still we have a knowledge of it that is nonetheless indiscriminate: for it could be any one of the other human beings. But if we add to this indiscriminate knowledge the individual properties of Socrates, we make it the most precise knowledge of Socrates. *It is exactly in this way* (οὕτω γοῦν), Aristotle says, that children too first have an indiscriminate knowledge (συγκεχυμένην γνῶσιν) of fathers and know them as men and not as fathers; hence, they think all men are fathers. But later when they can eventually attend to the properties of their father then they get hold of an articulated knowledge (διηρθρωμένην γνῶσιν) of him and so distinguish him from other men" (Philop. *In Phys.* 11.12–13. Trans. Osborne, with modifications).

Physics I.2

Timothy Clarke

The second and third chapters of *Physics* I contain an extensive critique of Eleatic monism, the theory of Parmenides and Melissus that 'what is is one'. In the second chapter Aristotle argues that this theory is impossible, and in the third chapter he explains why the Eleatics' arguments do not succeed.[1]

The second chapter can be divided into five sections:

1. An initial classification of the different theories of principles (184b15–25).
2. Prefatory remarks on the status of an examination of the Eleatics (184b25–185a20).
3. First sequence of criticisms of the Eleatics' monistic position, based on the claim that 'being is said in many ways' (185a20–b5).
4. Second sequence of criticisms of the Eleatics' monistic position, based on the claim that 'one is said in many ways' (185b5–25).
5. An excursus on some problems of one and many (185b25–186a3).

In what follows I shall discuss each of these sections in order, but my primary focus will be Aristotle's criticisms of the Eleatic position in the third and fourth sections. How exactly should we understand these criticisms, and what interpretation of the Eleatic position do they presuppose? I shall argue that we can make good sense of the criticisms when we see them as targeting two radical forms of monism: entity monism (the view that reality consists of just a single entity) and essence monism (the view that reality is all of the same essence).

I thank Verity Harte, Michael Della Rocca, Alan Code and Susanne Bobzien for detailed discussion of earlier drafts of this chapter. I am also grateful to the audiences at the workshops in Rome for their comments and suggestions, and to Francesco Ademollo, Laura Castelli, David Charles, Michel Crubellier, Ana Laura Edelhoff, Andrea Falcon, Lindsay Judson, James Lennox and Diana Quarantotto for further helpful conversation.

[1] For convenience I shall refer to Parmenides and Melissus together as 'the Eleatics'.

I. The Initial Classification of Theories (184b15–25)

Following the methodological preliminaries of *Phys.* I.1, the second chapter begins with a general classification of the different theories of principles:

> There must either be one principle or more than one. If there is one, it must either be unchanging, as Parmenides and Melissus say, or changing, as the natural philosophers say, some saying that the first principle is air, others that it is water. And if there is more than one, there must either be a limited or an unlimited number. If a limited number greater than one, there must either be two or three or four or some other number. And if there are an unlimited number, then either, as Democritus says, the genus is one and they are distinguished by shape or by species,[2] or they are also opposites. (184b15–22)

This classification serves to provide the framework for the book's subsequent investigation of principles. Aristotle first argues against the two most extreme positions mentioned here: the monistic theory of the Eleatics (I.2–3), and then the Anaxagorean theory that there are an unlimited number of principles, some of which are opposites (I.4). Once these two extreme positions have been refuted, he goes on to develop his own positive account of the number and nature of the principles, in *Phys.* I.5–7.

This sort of classificatory schema is not unique to Aristotle.[3] Something similar to it appears in Plato's *Sophist*, where the Eleatic Visitor describes various earlier theories of 'how many beings there are and what they are like' (242c4–243a1). Another related classification is given by Isocrates at *Antidosis* 268. One difference between Aristotle's classification and these others is that Aristotle's is a classification of theories of *principles*, whereas they are classifications of theories of *beings* or of *things that are* (τὰ ὄντα). In the next lines Aristotle suggests that this difference is merely terminological:

> And those who inquire into the number of things that are are also inquiring in a similar way. For they inquire primarily[4] into what the things that are are from, asking whether these things are one or many, and if they are many, whether there are a limited or unlimited number, so that they are

2 Reading σχήματι δὲ ἢ εἴδει διαφερούσας at 184b21–2 with E, I, Simplicius and Themistius. Cf. Crubellier (forthcoming).

3 Mansfeld 1986 gives an account of its origins.

4 Reading ζητοῦσι πρῶτον at 184b23 with F, I and J.

inquiring into whether the principles and the elements are one or many. (184b22–5)

The philosophers presented by Plato and Isocrates as inquiring into the number of 'things that are' are primarily inquiring into the number of *fundamental* entities, 'what the things that are are from'. So they are effectively inquiring into the number of principles. Aristotle does not say why he prefers to speak in terms of 'principles' rather than in terms of 'things that are', but this is presumably because the latter way of speaking is misleading, given that 'things that are' could easily be taken to refer to beings in general, rather than specifically to those that are fundamental.

It is worth noting that the classification at 184b15–22 does not represent Aristotle's final view of how best to catalogue the positions of his predecessors. He suggests revisions as the book goes on. For example, he will shortly argue that Parmenides and Melissus should not be seen as offering a theory of principles at all (I.2.185a3–5). And later on he will propose alternative ways of thinking of the theories of the material monists and of Democritus.[5]

II. Introduction to the Critique of the Eleatics (184b25–185a20)

The first of the various theories to be examined is the Eleatic theory that 'what is is one and unchanging' (184b26). In *Phys.* I.2–3 Aristotle concentrates almost exclusively on the first aspect of this theory, the claim that what is is one. He returns to the Eleatics' claim that what is is unchanging in I.8, when he addresses their argument against the possibility of coming to be. There is a good strategic reason for him to split up his treatment of the Eleatics in this way. His criticisms of their monism do not rely on his own theory of principles. By contrast, it is only when that theory of principles is on the table that we are able to see why a key Eleatic argument against change does not succeed.[6]

Before beginning his refutation of the Eleatics' monistic position, Aristotle makes some important prefatory remarks about the *status* of his critique of the Eleatics. The task of examining their theory is not, in fact, a task for the natural philosopher. This is because the Eleatics effectively

[5] See *Phys.* I.5.188a19–27, where he suggests that both the material monists and Democritus effectively make certain opposites principles – in the case of the material monists, the rare and the dense; in the case of Democritus, the full and the empty, as well as the various opposed forms of position, shape and arrangement.

[6] For the latter point see I.8.191a23–4 and 191b33–4. In Clarke 2015 I explain how Aristotle's theory of principles allows him to answer the Eleatic argument against change.

deny one of the basic presuppositions of the inquiry into the principles of nature – namely, *that there are principles*. (So their theory is not really a theory of principles at all, notwithstanding its inclusion in the initial classification.) The existence of a principle requires there to be some further entity (or entities) of which that principle is the principle; and yet this is excluded by the Eleatics' monism: 'For there is no longer any principle if it [sc. what is] is only one, and one in this way; for a principle is a principle *of* some thing or things' (185a3–5). This argument indicates that Aristotle takes Parmenides and Melissus to be claiming that there is only one (token) entity. This is the only form of monism that would rule out the existence of principles for the reason given here. I shall call this extreme form of monism 'entity monism'. It should be distinguished from other, less radical forms of monism such as 'substance monism', the view that there is only one (token) substance.[7]

If the task of examining the Eleatic theory does not fall to the natural philosopher, then whose task is it? Aristotle's language suggests that he thinks the task belongs to the dialectician.[8] Investigating the Eleatic position is not part of natural philosophy, but instead 'is like arguing dialectically (διαλέγεσθαι) against any other thesis put forward for the sake of argument . . . or like solving an eristic argument' (185a5–8).[9] Another possibility, favoured by some commentators, is that the task belongs to the higher science of metaphysics.[10] But the argument at 185a1–3 seems completely general: it does not fall to *the person investigating principles* (τῷ περὶ ἀρχῶν) to answer an opponent who denies the existence of principles. There is no indication that this is true only for some investigations of principles but not for others. And metaphysics, like natural philosophy, studies principles – the principles of being *qua* being. So it

[7] It is sometimes thought that Aristotle takes Parmenides' monism to be less extreme than Melissus'. For example, Palmer (2009) suggests that Aristotle interprets Melissus as an entity monist (in Palmer's terminology, a 'strict' monist), and Parmenides as a substance monist who allows for the existence of a plurality of non-substantial entities in addition to his one substance (a 'generous' monist). A disadvantage of Palmer's view is that it means that the present argument is a failure: if Parmenides allows for the existence of a plurality of entities, then his theory does not rule out the existence of principles, and accordingly Aristotle's argument fails to establish that an engagement with the Eleatic position falls beyond the scope of natural philosophy. A further difficulty is that Aristotle's subsequent criticisms of the Eleatic position do not target 'generous' substance monism, as we shall see below.

[8] Cf. also Irwin 1988: 67; Bolton 1991: 14–15; Falcon 2005: 28.

[9] Cf. also 185a19–20: 'it is presumably a good idea to have a little dialectical discussion (διαλεχθῆναι) about them [sc. the Eleatics]'.

[10] For this view see e.g. Wicksteed and Cornford 1934: 12; Mansion 1953: 172–3; Berti 1969.

would seem that if it is not the natural philosopher's job to answer the Eleatics, then nor is it the job of the metaphysician, for the same reason.[11]

Despite the fact that an examination of the Eleatic theory lies outside the scope of natural philosophy, Aristotle spends a lot of time on the Eleatics in *Physics* I. His justification is that 'although <the Eleatics do> not <speak> about nature (περὶ φύσεως), they nonetheless happen to state physical difficulties (φυσικὰς ἀπορίας)' (185a18–19).[12] The Eleatics do not speak about nature:[13] in rejecting the existence of plurality and change they are rejecting the existence of the natural world. Nevertheless, they do raise 'physical difficulties', that is, difficulties *relevant* to natural philosophy. The difficulties in question are presumably the puzzles about plurality and change that Aristotle will go on to address in the remainder of *Phys.* I.2–3 and in I.8. One way in which these Eleatic puzzles are relevant to natural philosophy is precisely that they call into question the most fundamental assumptions of this enterprise. There can be no science of nature if there are no principles, and there can be no nature if there is no change. Thus the Eleatics pose a major challenge to natural philosophy, and in Aristotle's view this challenge has yet to be properly answered.

Another, related, reason why the Eleatic puzzles are relevant to the inquiry into the principles of nature is that these puzzles greatly influenced post-Eleatic physics.[14] For example, Aristotle tells us that the atomic theory of Leucippus and Democritus arose out of reflection on Eleatic arguments (see *GC* I.8.324b35–325b5; *Phys.* I.3.187a1–3). The atomists posited atoms and void in an attempt to show how, contrary to the Eleatics, the world can be plural and changing. This means that our evaluation of atomism as a theory of the principles of nature – our stance on whether or not it is well founded – will depend in part on our own view of how best to resolve the Eleatic puzzles that motivate it.

[11] I take this to be consistent with the fact that Aristotle's subsequent refutation of the Eleatics is 'metaphysical' in the sense that it relies on various metaphysical claims, such as the claims that being and one are said in many ways.

[12] Following Ross's punctuation: περὶ φύσεως μὲν οὔ, φυσικὰς δὲ ἀπορίας. Alternatively, one could place the comma after μὲν, in which case Aristotle would be justifying his discussion of the Eleatics on the grounds that 'they happen to speak about nature, although they state difficulties that are not physical'. See Simpl. *In Phys.* 70.5–17. The latter reading is preferred by A. Mansion (1945: 66 n. 48) and Pellegrin (1994: 126–8), among others.

[13] It is plausible that Aristotle is here taking issue with the standard titles (or descriptions) of the works of Parmenides and Melissus. While Parmenides' poem is unlikely to have had a title originally, at some point it came to be known as a work *On Nature* (Περὶ φύσεως). The disjunctive title ascribed to Melissus' book by Simplicius – *On Nature, or On What Is* (Περὶ φύσεως ἢ περὶ τοῦ ὄντος) – was probably Melissus' own: see Palmer 2009: 205–6 n. 25 and Harriman 2015: 19–20.

[14] Cf. also Laura Castelli's contribution to this volume.

For at least these reasons, then, the Eleatic challenge is highly relevant to natural philosophy, and in particular to the inquiry into the principles of natural beings. This accounts for Aristotle's decision to include a critique of the Eleatics in the opening book of the *Physics*, even though the critique falls beyond the scope of natural philosophy itself.

III. The First Sequence of Criticisms (185a20–b5)

Aristotle now begins his refutation of the Eleatic doctrine that 'what is is one'.[15] The first sequence of criticisms consists of two arguments. In the first Aristotle uses the theory of categories to construct a dilemma for the Eleatics (185a20–32). In the second he raises a problem for Melissus' claim that what is is unlimited (185a32–b5).

III.1 Being is Said in Many Ways

The first argument runs as follows:

> Since being (τὸ ὄν) is said in many ways, the most appropriate starting-point (ἀρχή) of all is to ask in what way those who say that 'all things are one' speak <of being> – whether all things are substance, or quantities, or qualities, and again whether all things are one substance, like one human being, or one horse, or one soul, or whether all things are quality, and this is one, like pale or hot or one of the other things of this sort. For all these differ a great deal, and all are impossible to maintain. For if, on the one hand, there is substance and quality and quantity, then whether these things are detached from one another or not, the things that are will be many. But if, on the other hand, all things are quality or quantity, then whether substance is or is not, this is absurd, if one should call the impossible absurd. For none of the others is separate apart from substance. For all <the others> are said of substance as an underlying thing. (185a20–32)

According to a common reading of this passage, Aristotle starts from the claim that the word 'being' is ambiguous ('being is said in many ways'). There are many different senses of 'being', one for each of the different categories: substance, quantity, quality, and so on. And so there are

[15] Alternative formulations of the doctrine are: 'the universe is one' (εἶναι ἓν τὸ πᾶν, 185b7); 'all things are one' (εἶναι ἓν τὰ πάντα, 185a22); 'the beings are one' or 'the things that are are one' (ἓν εἶναι τὰ ὄντα, 185b24); and 'all beings are one' or 'everything that is is one' (ἓν τὰ ὄντα πάντα, 185b19–20). Aristotle evidently regards these different formulations as equivalent, and moves freely between them.

different things that the Eleatics might mean when they say that '*what is* is one' or that 'all *beings* are one'. Aristotle uses the doctrine of the categories to disambiguate the Eleatic thesis, and argues that on each of the possible disambiguations the Eleatics are committed to absurdity.[16]

This reading of the passage is tempting, but it faces a serious difficulty. If it were correct, Aristotle ought to consider the following as possible disambiguations of 'all beings are one': (A) all *substances* are one; (B) all *qualities* are one; (C) all *quantities* are one, and so on. However, he does not engage with any of these positions. His subsequent argument rests on what I shall call the 'interdependence thesis':

Substances depend for their existence on the existence of non-substances, while non-substances depend for their existence on the existence of substances.[17]

It is hard to see how this thesis could be thought to undermine any of the monistic positions just mentioned. Take claim (A), 'all substances are one'. This presumably means either (A1) that there is just one (token) substance, or (A2) that all substances are of a single type. But both of these positions are consistent with the above thesis. Even if substances cannot exist without non-substances, and vice versa, it might still be the case that there is only one token substance, or only one type of substance.

An alternative reading is therefore required – a reading which can accommodate the fact that the argument proceeds by appeal to the interdependence thesis. I suggest that it is a mistake to take Aristotle to be claiming that 'being' is ambiguous. As others have noted, when he says that *F* 'is said in many ways' (πολλαχῶς λέγεται), it is sometimes best to interpret him as meaning not that the word '*F*' has many senses, but instead that there are many kinds of *F*-ness, or many ways of being *F*.[18] His claim at 185a21 that 'being is said in many ways' is a case in point. We should take him to mean that there are many different ways of being:[19] substances have a different way of being from qualities; qualities have a

[16] For this interpretation see e.g. Ross 1936: 338, 467; Mansion 1953: 173; Gershenson and Greenberg 1962: 139–40; Palmer 2004: 49.

[17] When I say that '*X* depends for its existence on the existence of *Y*', I mean that *X* cannot exist unless *Y* also exists.

[18] See e.g. Barnes 1995: 73–5. As Barnes notes, it is possible for two things to be *F* in different ways without its being the case that '*F*' has different senses as applied to each of them: 'it is one thing for an argument to be good and quite another for, say, a dinner to be good; and yet the word "good" is not ambiguous between cases of this sort' (75 n. 7). On the philosophical importance of distinguishing between senses of '*F*' and ways of being *F* (or kinds of *F*-ness), see Matthews 1972.

[19] My use of the terminology of 'ways of being' follows that of Frede (1987: 85).

different way of being from quantities, and so on.[20] His question, then, is about the way of being that the Eleatics want to attribute to what is. He uses the doctrine of the categories to distinguish various possible answers that they might give to this question, and then argues that each of these answers has absurd consequences. As we shall see, this reading fits much better with the fact that the argument relies on the interdependence thesis.

III.2 The Argument's Targets

I have denied that at 185a20–32 Aristotle is using the doctrine of the categories to disambiguate the Eleatics' claim that 'what is is one'. But then how exactly does he understand their position? What kind, or kinds, of monism is he arguing against in this passage? When earlier he argued that the Eleatics' theory is incompatible with the existence of principles, his reason was that there cannot be any principles at all if, as the Eleatics claim, there is only one entity (185a3–5). We should therefore expect entity monism to be a target of the present argument. And I think it is; but it does not seem to be the only target.

After saying that we must ask 'in what way those who say that "all things are one" speak <of being>', Aristotle then raises two more specific sub-questions. First: do they hold that all things are substance, or quantities, or qualities, and so on? Second: do they hold that all things are one substance, or one quality, and so on?

How should we understand the first sub-question? I think that we can set aside Ross's interpretation, according to which the question concerns the meaning of 'all things' (or 'all beings') as it occurs in the Eleatic claim: does it mean 'all substances', or 'all quantities', or 'all qualities'?[21] It is unlikely that this is Aristotle's question, for he does not go on to argue against the claims that 'all substances are one', 'all quantities are one', and so on.

But if we do not read the question as a question about the meaning of 'all things', then how should we read it? It asks whether, according to the

[20] One might worry that Aristotle is begging the question against the Eleatics in claiming that *there are many* categories and that *there are many* ways of being. These claims might seem to presuppose the falsity of the Eleatics' radical monism. But, in response to this, Aristotle's position can be understood as being that (1) for any entity, *X*, *X* is either a substance, or a quantity, or a quality (and so on), and that (2) anything that is a member of one category has a different way of being from anything that is a member of another category. Both of these claims are free of any existential commitments, and so neither presupposes the falsity of the Eleatics' monism.

[21] Ross 1936: 467.

Eleatics, all things are substance or *quantities* (ποσά) or *qualities* (ποιά). The plurals apparently indicate that Aristotle has in his sights a version of monism that allows for the existence of a plurality of entities of a single kind. Thus, the first sub-question suggests that entity monism is not his sole target. If it were, the question would make little sense. Entity monism is obviously incompatible with the existence of multiple quantities or of multiple qualities. The second sub-question would still of course be a relevant question (are all things one substance, or one quality . . . ?), but the first would not.

I therefore want to suggest that Aristotle also means to argue against 'essence monism', the view that all of reality ('what is') is of the same essence, or the same nature. This is a view which, considered by itself, allows for the existence of a plurality of entities.[22] That Aristotle considers essence monism an authentically Eleatic doctrine is suggested by another of his criticisms later in the chapter, where he argues, against the Eleatics, that essence monism leads to unacceptable 'Heraclitean' consequences (see 185b19–25, to be discussed below). His attribution of essence monism to Parmenides is confirmed by an argument in I.3, where he criticises Parmenides for failing to see that he is unable to establish essence monism (186a28–32). This criticism presupposes that Parmenides was *trying* to establish essence monism. There is also a passage in Aristotle's examination of Melissus' argument which suggests that he attributes an equivalent position to Melissus (namely, that what is is one 'in form': see 186a19–22). So, I suggest that the first sub-question is asked with essence monism in mind. If all of reality is of the same essence, then what way of being do things have? Are they substances, or quantities, or qualities (and so on)? An essence monist must choose one of these options. There cannot be substances *and* quantities, for instance, because then it would not be the case that all of reality is of the same essence. (If item *X* belongs to one category and item *Y* to another, then *X* and *Y* have different essences.)

The second sub-question, by contrast, is asked with *entity* monism in mind. If reality consists just of one token entity, what way of being does it have? Is it one substance (like a single human being), or one quality (like paleness), or a single item in one of the other non-substance categories?

[22] This is not to say that a *proponent* of essence monism will necessarily allow for the existence of a plurality of entities. It is possible that someone might claim both that (1) reality consists of just a single entity (entity monism) and that (2) reality is all of the same essence (essence monism). In my view, Aristotle attributes this conjunctive position both to Parmenides and to Melissus.

My suggestion, then, is that Aristotle is arguing in the present passage against both of these varieties of monism: entity monism and essence monism. This explains the two sub-questions at 185a22–6 – a detail that we could not adequately explain if he were targeting entity monism alone.

III.3 The Consequences of Interdependence

We now come to Aristotle's use of the interdependence thesis. As we have just seen, an essence monist is committed to saying that either

(i) reality consists only of substances;
or
(ii) reality consists only of entities of a single non-substance kind (either qualities, or quantities, or relatives, and so on).

An entity monist, on the other hand, is committed to saying that either

(iii) reality consists only of a single substance;
or
(iv) reality consists only of a single non-substantial entity (either a quantity, or a quality, or some other non-substantial entity).

None of these options is possible – or as Aristotle puts it at 185a26–7, 'all are impossible to maintain'. The reason is that they all conflict with the interdependence thesis.

Aristotle first argues that it cannot be the case either that (i) reality consists only of substances, or that (iii) it consists only of a single substance.[23] At any rate, I take this to be the point of the following sentence:

> For if, on the one hand, there is substance and quality and quantity, then whether these things are detached from one another or not, the things that are will be many. (185a27–9)

It is not immediately obvious, of course, that this sentence is meant to explain why reality cannot consist only of substances or only of a single substance. After all, the claim here is that if there is substance *and* quality *and* quantity, then the things that are will be many, and not, as the Eleatics claim, only one. Nevertheless, it seems to me likely that Aristotle's intention here is to argue against claims (i) and (iii). I suggest that he is

[23] It is not that he gives two separate arguments, first an argument for why (i) is impossible, and then an argument for why (iii) is impossible. Rather, the same argument is intended to rule out both (i) and (iii) at the same time.

effectively relying on the first part of the interdependence thesis: substances depend for their existence on the existence of non-substances. Any substance will need to possess at least *some* non-substantial attributes; for example, if there is a human being, this human being must have certain features: it must have various qualities, be of a certain size, and so on. Since there can be no substances that are completely bereft of non-substantial attributes, to commit oneself to the existence of a substance is thereby to commit oneself to the existence of other, non-substantial entities as well. The fact that these non-substantial entities are not 'detached' from (but rather are inherent in) the substance is irrelevant: they are additional entities all the same. It follows that reality cannot consist only of substances, or only of a single substance.

According to this reading, the point being made at 185a27–9 is really this: if the Eleatics accept the existence of substance, then – because there can be no substances without non-substantial attributes – there will be substance *and* quality *and* quantity (for example). And if substance, quality and quantity all exist, then what is will be many: reality will consist of many entities (*contra* entity monism), and of many kinds of entity (*contra* essence monism). And this will be so even if these entities are not 'detached' from one another.

This is admittedly reading quite a bit into the text at 185a27–9, but doing so seems warranted for the following reason. If Aristotle were not here arguing against the possibility that reality consists only of substances or only of a single substance, we would be given no argument against these claims. Given that they were among the options mentioned earlier on (at 185a22–4) and said to be impossible (at 185a26–7), there would then be a puzzling lacuna in Aristotle's case against the Eleatics. This seems to be a strong consideration in favour of the proposed reading of the present sentence.

He next argues that it cannot be the case either that (ii) reality consists only of entities of a single non-substantial kind, or that (iv) it consists only of a single (token) non-substantial entity:

> But if, on the other hand, all things are quality or quantity, then whether substance is or is not, this is absurd, if one should call the impossible absurd. For none of the others is separate apart from substance. For all <the others> are said of substance as an underlying thing. (185a29–32)

This is the second part of the interdependence thesis: non-substances depend for their existence on the existence of substances. Quantities and qualities cannot exist on their own, without anything of which they are the quantities and qualities. Thus reality cannot consist only of quantities or of a single quantity (for example). Aristotle says that the claim that all things are quantity or quality is absurd 'whether substance is or is not' (185a29–30). The idea is that if the Eleatics hold that all things are quantities or a single quantity, and yet admit the existence of substance, then they are contradicting themselves (because if substances exist, then it is not the case that all things are quantities or a single quantity). If, on the other hand, they deny that substances exist, then they are committed to the existence of free-floating attributes, which is absurd.

It follows that neither essence monism nor entity monism can be true. By the interdependence thesis, if the world contains anything at all, it must contain both substances and non-substantial attributes. So it cannot be the case that there is only one entity, or that all of reality is of the same essence. Whatever way of being the Eleatics want to attribute to what is, both of these varieties of monism are impossible.

III.4 Against Melissus' Unlimited One

The second argument of the first sequence is directed against Melissus, and focuses on his claim that what is is unlimited (a claim Melissus argues for in B2–4). Melissus takes the unlimitedness of what is to entail entity monism, the thought being that if anything that exists is unlimited in extent, there can be no room for more than one thing (B6). In response, Aristotle argues that the unlimitedness of what is actually entails entity *pluralism*:

> And Melissus says that what is is unlimited. Therefore what is is a quantity (ποσόν). For the unlimited is in the <category of> quantity, and it is not possible for a substance or a quality or an affection to be unlimited, except incidentally, if they are at the same time also certain quantities. For the account of the unlimited employs quantity, but not substance or quality. If, therefore, it is both a substance and a quantity, what is is two and not one. But if it is substance alone, then it is not unlimited, nor even will it have any magnitude at all. For then it will be a quantity. (185a32–b5)

Aristotle's first move is to argue that Melissus' claim that what is is unlimited entails that what is is a quantity (ποσόν), that is, the bearer of a quantitative property.[24] This is not absurd in itself, of course; the problem

[24] The word ποσόν, which I am translating as 'quantity', is ambiguous between a quantitative *property* (e.g. the property of being six feet tall) and the *bearer* of such a property (e.g. a person six feet tall). I take it that the latter is what is intended here. Cf. also *Cat.* 6 for these two uses of ποσόν.

arises from the fact that Melissus also claims to be a monist. If what is is unlimited, and so the bearer of a quantitative property, then it cannot be that there is just one entity. There will be at least two entities: the quantitative property, unlimitedness, and the thing that has it, a substance. If, on the other hand, only the substance were to exist, bereft of any quantitative properties, the Melissan One would not have any size at all.

It seems reasonable to think that Aristotle intends this argument against Melissus to serve as a specific instance of a more general objection, namely, that the Eleatics' attribution of various properties to their one being is incompatible with their monism. We might speculate that Aristotle singles out the Melissan property of unlimitedness because he likes the particular irony of the example. This is a property which Melissus explicitly says entails monism, but which in fact entails its negation.

The Eleatics themselves apparently saw no tension between their monism, on the one hand, and their attribution of many different properties to what is, on the other. A plausible explanation of this is that they considered only the *underlying bearers* of properties to be beings, and not the properties themselves. Aristotle can thus be seen as highlighting (what he sees as) the Eleatics' naive and overly restrictive conception of what counts as a being. In his view properties are beings too. The significance of this fact extends beyond the critique of Eleatic monism, for later in *Physics* I he will develop a theory of principles – the theory of substratum, form and privation – which requires that we recognize properties as well as property-bearers as entities within our ontology. (The underlying substratum is a property-bearer; the form and the privation are properties predicated of the substratum.) Aristotle's critique of monism takes the Eleatics to task for overlooking the existence of properties, and in doing so prepares the way for the positive account of principles that is to follow.

IV. The Second Sequence of Criticisms (185b5–25)

Aristotle's second sequence of criticisms begins from the claim that one, like being, is said in many ways:

> Further, since the one itself is also said in many ways, just as being is, it is necessary to investigate in what way they say that the universe is one. And we call one either the continuous, or the indivisible, or those things of which the account of their essence is one and the same, such as μέθυ and οἶνος.[25] (185b5–9)

[25] These are two words for wine.

I argued above that the claim that 'being is said in many ways' at 185a21 should be understood as a claim about ways of being, and not about senses of the word 'being'. Similarly for the claim that 'the one itself is also said in many ways': we should take Aristotle to be distinguishing three different ways of being one, rather than three different senses of the word 'one'.[26]

To see this, consider the first item on the list: *continuous.* If Aristotle were suggesting that this is a possible sense of 'one', he would effectively be suggesting that the Eleatics' claim that 'what is is one' might be interpreted as the claim that 'what is is continuous'. But in that case the objection that follows would fail. The objection is that if what is is continuous, then 'the one is many': the Eleatics' one being will consist of infinitely many parts (185b9–11). But if the Eleatics' monistic thesis *just is* the thesis that 'what is is continuous', this consequence is hardly problematic. A thing's having infinitely many parts is obviously consistent with its being continuous. So if 'continuous' is intended to be a candidate for what the Eleatics *mean* by 'one', Aristotle has failed to come up with a good objection to their position.

In order to rescue the argument we should take Aristotle to be suggesting that continuity is a *way* in which what is might be one. Some objects are one by being continuous: what makes them one object, rather than many, is that each of their parts shares a boundary or a limit with some other part.[27] But suppose that the Eleatics are entity monists. Then we can easily understand why (in Aristotle's view) they cannot consistently say that what is is one in this way. If the Eleatics' one entity is continuous, then it is divisible into infinitely many parts, in which case there are many entities, not one. The 'way of being one' reading therefore gives Aristotle a good objection, whereas the 'sense of "one"' reading does not.

It is best, then, to take Aristotle to be distinguishing three different options about the way in which what is is one: (1) what is is one by being continuous, (2) what is is one by being indivisible, and (3) what is is one in account. This is not an exhaustive list of ways of being one; further ways are distinguished in the discussions of oneness in *Metaph.* Δ 6 and I 1. Rather, the three options are selected for their relevance to the discussion of the Eleatics. Each is a plausible candidate for what the Eleatics might have had in mind as to the way (or *a* way) in which what is is one.

[26] *Pace* e.g. Wicksteed and Cornford (1934: 23–5); Cherniss (1935: 63–4); Ross (1936: 338); Gershenson and Greenberg (1962: 140); Stokes (1971: 1); Palmer (2004: 49); Bostock (2006b: 103); Coxon (2009: 33).

[27] For this understanding of continuity, see *Phys.* V.3.227a10–12.

IV.1 Continuity

We can now look more closely at Aristotle's arguments against each of the three options. The first option, as we have seen, is that what is is one by being continuous. Parmenides explicitly claims that what is is continuous at B8.6,[28] and he argues for this claim at B8.22–5:

> Nor is it divided, since it is all alike;
> nor is it any more here, which would prevent it from holding together,
> nor any less, but it is all full of being;
> thus it is all continuous, for being draws near to being.[29]

On what I take to be the most straightforward interpretation of this passage, it argues that what is is a spatially continuous object (among other things). There are no gaps between regions of being ('it is all full of being'); therefore what is is spatially continuous.[30] It is plausible that this is at least part of Parmenides' reason for claiming that what is is 'one' (B8.6), given that anything continuous is thereby a unity.[31]

Aristotle's objection to the claim that what is is one by being continuous is based on the principle that 'the continuous is divisible to infinity' (εἰς ἄπειρον γὰρ διαιρετὸν τὸ συνεχές).[32] It may be worth clarifying his notion of divisibility here. I take it that he does not mean that continuous things are *physically* divisible to infinity, where 'physical division' refers to a process whereby formerly adjoining parts of a thing come to be separated by a spatial interval.[33] This does not seem to be what he intends, since he is making a general claim about continuous things, and in his view there are continuous things that are not subject to physical division, such as periods of time. I suggest that for something to be 'divisible' in the relevant sense it is enough if different parts of it can be distinguished from one another, regardless of whether they can be

[28] Assuming that the orthodox reading of the start of the line – ἕν, συνεχές ('one, continuous') – is correct.

[29] οὐδὲ διαιρετόν ἐστιν, ἐπεὶ πᾶν ἐστιν ὁμοῖον· | οὐδέ τι τῇ μᾶλλον, τό κεν εἴργοι μιν συνέχεσθαι, | οὐδέ τι χειρότερον, πᾶν δ' ἔμπλεόν ἐστιν ἐόντος. | τῷ ξυνεχὲς πᾶν ἐστιν· ἐὸν γὰρ ἐόντι πελάζει.

[30] Some think that Parmenides intends a temporal rather than a spatial sense of 'continuous' (see e.g. Owen 1960: 97). However, the spatial interpretation is strongly suggested by the use of the spatial adverb 'here' (τῇ) at line 23, and by the verb 'draws near' (πελάζει) at line 25.

[31] Cf. also Barnes 1979: 11: 'When Parmenides juxtaposes "ἕν" ["one"] and "συνεχές" ["continuous"], it is only reasonable to suppose that the second word is intended to explicate the first'.

[32] Aristotle sometimes treats this as a *definition* of 'continuous': see e.g. *Phys.* VI.2.232b24–5 and *DC* I.1.268a7–8; cf. also *Phys.* III.1.200b16–20.

[33] I take this definition of 'physical division' from Furley 1967: 4.

physically separated. The claim, then, is that for any continuous thing, *C*, it is possible to distinguish different parts of *C*, and of each of these different parts it is possible to distinguish further sub-parts, and so on *ad infinitum*.[34]

Must an Eleatic agree that it is possible to distinguish multiple parts of any continuous being? Parmenides may be thought to deny this when he says, at the beginning of the passage quoted above, that 'Nor is it [sc. what is] divided (διαιρετόν), since it is all alike' (B8.22). The argument appears to be that, because what is is uniform, it is not divided into parts. But to this Aristotle might reasonably reply that if something is continuous then it *must* contain parts, even if it is perfectly uniform. Something continuous (συνεχές) is literally something that 'holds together' (συνέχεται). But it is hard to know what it could mean for a thing to 'hold together' if not that it has *parts* that 'hold together' *with one another*. And once the Eleatics have conceded this, they are in trouble. Or rather, they are in trouble if their position is that reality consists of only one entity.[35]

IV.2 Indivisibility

What is cannot be one by being continuous, at least not if the Eleatics are to be consistent entity monists. The second option that Aristotle considers is that what is is one by being indivisible. I take him to be putting this

[34] This allows us to respond on Aristotle's behalf to a problem raised by Bostock (2006b: 104). Bostock worries that Aristotle's own doctrine of parts is that the parts of a whole exist only in so far as they are capable of being separated from that whole. Since the Eleatics will claim that the putative parts of their one entity are *incapable* of being separated, they therefore have Aristotelian grounds on which to resist the present objection. In response to this worry, Aristotle can say that even if the Eleatic One lacks separable parts, it is nonetheless the case that different parts of it can be distinguished from one another.

[35] Immediately after arguing that the continuity of what is would contravene the Eleatics' (entity) monism, Aristotle mentions a difficulty about parts and wholes (185b11–16). The basic difficulty is that of the relation between a whole and one of its parts – a pen and its nib, say: are they one thing or two? He does not explain why he mentions this difficulty here, but I suggest that the point is that if the whole and the part were simply identical, his foregoing objection would not succeed. The divisibility of what is into parts would not add to the total number of entities in existence, and so would not threaten entity monism. Now, Aristotle also says that this difficulty is not πρὸς τὸν λόγον but is rather αὐτὴν καθ' αὑτήν (185b12). But I take it that he is not saying that the difficulty is *irrelevant to* his argument against the Eleatics, contrary to the usual view of interpreters. (If so, why mention it?) Rather, he is saying that this difficulty is not a difficulty *for his argument*, even though it nonetheless is a difficulty *in its own right*. The reason why the difficulty is not a difficulty for Aristotle's argument is that, however it is to be resolved, it is obvious that the correct answer will *not* be to say that the relation between a whole and one of its parts is that of identity. That answer is an obvious non-starter, and yet it is the only possible answer that could threaten Aristotle's foregoing argument against the Eleatics.

forward as a second way in which the Eleatic entity monist's one entity might be one. The previous criticism was that anything that is one by being continuous must be divisible into parts, and the existence of multiple parts is incompatible with entity monism. So it would seem that a better option for the Eleatic entity monist is to abandon the claim that what is is one by being continuous, and to say instead that it is one simply by virtue of its being indivisible into parts.

As before, the relevant notion of divisibility here is not physical divisibility. If the Eleatics are to avoid the problem raised against the first option, it is not enough for them to say that what is is physically indivisible, because physical indivisibility is compatible with having multiple parts. Rather, to avoid the objection they must say that what is is indivisible in the sense of being absolutely partless: a mereological atom. As noted above, Parmenides may be claiming that what is is indivisible in this sense at B8.22. Melissus also appears to have rejected the existence of parts, although this is controversial.[36]

Aristotle's objection to this second option, unlike his objection to the first, is not that it contravenes entity monism. The objection this time is that the view conflicts with other Eleatic commitments: 'But if it is one by being indivisible, then nothing will be a quantity or a quality, nor then will what is be unlimited, as Melissus says, nor limited, as Parmenides says. For the limit is indivisible, but not the limited thing' (185b16–19).

The underlying assumption here is evidently that a thing can be extended only if it has parts. Given this assumption, an indivisible or partless Eleatic One must be unextended, like a geometrical point. It will therefore lack any quantity or size. Yet on Aristotle's interpretation both Parmenides and Melissus hold that their One has size.[37] So if they want to claim that what is is indivisible in the sense of being absolutely partless, they will be forced to abandon this view.

A little strangely, Aristotle also suggests that if what is is indivisible then it will lack any qualitative properties. This is odd because it does not in general seem true that indivisible things lack qualities.[38] The probable explanation is that he is thinking of the kinds of qualities that

[36] See Simpl. *In Phys.* 87.6–7 (= the second half of B9): 'being one, he [sc. Melissus] says, it must not have body; but if it had thickness, it would have parts, and no longer be one'. The attribution of this argument to Melissus has been questioned on the grounds that he should not want to claim that his infinitely large, plenistic One lacks 'body' and 'thickness'. For discussion see Palmer 2003.

[37] See the objection to Melissus at 185a32–b5, and the objection to Parmenides at I.3.186b12–14.

[38] Consider the first mover as described in *Metaph.* Λ 7. It is indivisible in the sense of being partless (1073a5–7), and yet it is also supremely good (1072b29), a quality.

the Eleatics want to attribute to their One, such as, in Parmenides' case, its sphericity.[39] If the One is indivisible, then it will lack size, and if it lacks size, it will lack the sorts of qualities the Eleatics want.

The claim that what is is indivisible (in the sense of being partless) is also inconsistent with Melissus' claim that it is unlimited, and with Parmenides' claim that it is limited. This is again for the reason that a thing cannot have such properties unless it is extended, and it cannot be extended unless it has parts.

IV.3 One in Account

The third and last option is that what is is one in account. For *X* and *Y* to be 'one in account' (in the relevant sense) is for the account (or definition) of the essence of *X* to be the same as the account (or definition) of the essence of *Y*. In other words, *X* and *Y* are one in account just in case they have the same essence. So, to say that 'what is is one in account' is to say that *everything that is* has the same account and the same essence. This is the position I earlier called 'essence monism', the position that all of reality is of the same essence.

As with the preceding two options, this third option too has a basis in the Eleatic texts. While there are variant readings, scholars now generally agree that at B8.4 Parmenides claims that what is is μουνογενές.[40] One natural way (if not the only way) of interpreting him here is as claiming that what is is 'of one kind'.[41] And a plausible restatement of this, in Aristotelian terminology, is that all of reality is of the same essence, or (equivalently) that everything that is is one in account. We might see further evidence of the Eleatics' commitment to essence monism in their claim that what is is all alike (see Parmenides B8.22, Melissus B7.1 and *MXG* I.974a12–14).

The objection to this final option runs as follows:

> But if everything that is is one in account, as are mantle and cloak, then it follows that they are affirming the account of Heraclitus. For being good and being bad will be the same, and being good and being not-good, so that the same thing will be good and not-good, and a human and a horse, and

[39] Aristotle nowhere explicitly says that the Parmenidean One is spherical, but it seems likely that this is his interpretation of B8.42–4 given that he seems to take literally Parmenides' description of what is as a spatially finite object 'equally balanced from the middle' (μεσσόθεν ἰσοπαλές): see *Phys.* III.6.207a15–17.

[40] In favour of this reading, see the arguments of Tarán 1965: 88–93.

[41] For this translation of μουνογενές, cf. Mourelatos 2008: 113–14; Palmer 2009: 140 n. 7.

> their account will not be about the fact that the things that are are one, but about the fact that they are nothing. And being this quality and being this quantity will be the same.[42] (185b19–25)

On the face of it, this objection is puzzling. The objection appears to assume that the claim that 'everything that is is one in account' entails, for example, that the properties good and bad have the same essence. But surely this is not a consequence of that claim. Of course, if one agrees that there *are* such properties as good and bad, the claim that 'everything that is is one in account' entails that they have the same essence. But why think that an Eleatic will agree to the existence of such things?

My suggestion is that Aristotle is implicitly attributing to the Eleatics the view that *it is impossible to speak or think of what is not.*[43] This view commits one to saying that, when we speak of the properties of good and bad, the referents of our terms 'good' and 'bad' exist (because one cannot speak of what is not). Hence, if everything that is is one in account, it follows that the referents of 'good' and 'bad' are one in account. Similarly, the view commits one to saying that when we *think* of good and bad, we are thinking, in each case, of something that is (because one cannot think of what is not). Hence, if everything that is is one in account, it follows that the properties we are thinking of have the same essence.

Aristotle's criticism of the third option now becomes rather more intelligible. If an essence monist accepts that it is impossible to speak or think of what is not, then it seems that they must agree that the essence of the property we call or think of as 'good' is the same as the essence of the property we call or think of as 'bad'. And this seems to entail that the properties good and bad are identical. Thus there is reason to think that essence monism commits the Eleatics to the 'Heraclitean' consequence that supposedly opposite properties are identical, with the result that anything that has the property *F* also has the property not-*F*.

This is obviously a deeply problematic result for the Eleatics – especially given how adamant they are that nothing is both *F* and not-*F*.[44] If nothing is both *F* and not-*F*, *and* anything *F* is also not-*F*, then it follows that nothing at all exists. (At least, this follows given the uncontroversial

[42] I repunctuate Ross's text as follows: ταὐτὸν γὰρ ἔσται ἀγαθῷ καὶ κακῷ εἶναι, καὶ ἀγαθῷ καὶ μὴ ἀγαθῷ εἶναι, ὥστε ταὐτὸν ἔσται ἀγαθὸν καὶ οὐκ ἀγαθόν, καὶ ἄνθρωπος καὶ ἵππος, καὶ οὐ περὶ τοῦ ἓν εἶναι τὰ ὄντα ὁ λόγος ἔσται ἀλλὰ περὶ τοῦ μηδέν. καὶ τὸ τοιῳδὶ εἶναι καὶ τοσῳδὶ ταὐτόν.

[43] Many commentators take this view to be expressed in Parmenides B2.7–8: 'For you could not apprehend what is not, for that is impossible, | nor indicate it' (οὔτε γὰρ ἂν γνοίης τό γε μὴ ἐόν, οὐ γὰρ ἀνυστόν, | οὔτε φράσαις).

[44] See esp. Parmenides B7.1: 'For never shall this prevail, that things that are not are.'

assumption that a thing cannot exist without some predicate, *F*, belonging to it.) This, I propose, is why Aristotle claims that, if everything that is is one in account, the Eleatics' account 'will not be about the fact that the things that are are one, but about the fact that they are nothing' (185b23–5). It turns out that the Eleatics are committed not to monism but to nihilism, the (by their own lights) unthinkable and unspeakable thesis that nothing exists at all.

V. Problems of One and Many (185b25–186a3)

We have now considered each of Aristotle's criticisms of Eleatic monism in *Phys.* I.2. I have suggested that he attributes two kinds of monism to the Eleatics, entity monism and essence monism, and that he argues in this chapter against both kinds. His next main task, taken up in I.3, is to explain why the Eleatics' arguments for their theory are unsuccessful. But first, in the final section of the present chapter, he includes a brief excursus on some later philosophers and their struggles with some problems of one and many:

> And the more recent of the early thinkers were also troubled lest the same thing should turn out for them to be at the same time both one and many. This is why some people took away the 'is', like Lycophron, while others refashioned their speech, saying not that the human 'is pale', but that he 'has-paled', and not that he 'is walking', but that he 'walks', so that they would never make the one be many by adding the 'is', supposing that the one or being is said in only one way. But the things that are are many, either in account[45] (for example, being pale and being educated are different, yet the same thing is both; therefore the one is many), or by division (as with the whole and the parts). But here they were already in difficulty, and they conceded that the one is many, as though it were not possible for the same thing to be both one and many – although not those that are opposed. For the one is both potentially and actually. (185b25–186a3)

The opening sentence of this passage ('the more recent of the early thinkers were *also* troubled ...') implies that the Eleatics were themselves troubled by the possibility that the same thing might be both one and many. Aristotle would seem to be thinking here primarily of Melissus, who argues that since what is has been shown to be one, it cannot have any feature that would cause it to be many, apparently assuming that it is

[45] A thing is 'many in account' (πολλὰ λόγῳ) in this sense just in case it has many properties (pale, educated and so on) the accounts of which are non-identical. This use of 'in account' should be distinguished from the use we just met at *Phys.* I.2.185b19–20.

impossible for a thing (in this case, what is itself) to be one and many at the same time.[46]

The more recent thinkers mentioned at 185b25–186a3 are not Eleatic monists, but they share the Eleatic concern about the possibility of the same thing's being both one and many. This concern motivates their proposed linguistic reforms: 'some people took away the "is" ... while others refashioned their speech'. These reforms are intended to avoid a problem that arises when we say that one thing, such as Socrates, is human, is pale, is educated, and so on. This ordinary way of speaking has the supposedly paradoxical consequence that *one thing*, Socrates, *is many different things* (human, pale, educated).[47] In an effort to avoid this problem, the more recent thinkers proscribed the use of the verb 'is' in sentences such as 'Socrates is pale'. Some of them, like the sophist Lycophron, proposed dropping the 'is' and using instead the verb-free formulation 'Socrates pale'. Others, whom Aristotle does not name, 'refashioned their speech', replacing the copula-plus-predicate phrase 'is pale' with the single verb 'has-paled'. These two reforms have a similar effect: they allow us to escape the seemingly contradictory conclusion that the one is many, by giving us alternative ways of saying that Socrates has the properties of paleness, educatedness, and so on. If we can express these claims without actually saying that Socrates '*is* pale' and '*is* educated', then it seems that we can avoid the problematic result that one thing *is* many different things.

The same thinkers were also troubled by a second one–many problem: that wholes of parts appear to be both one and many at the same time. Take Socrates again: he appears to be both one thing (a single human being) and also many different things (his head, torso, arms and legs).[48] Aristotle suggests that the more recent thinkers were unable to find a solution to this second problem, and so were here forced to accept the (to them) unwelcome conclusion that the one is many.

[46] See e.g. the argument for uniformity reported at *MXG* I.974a12–14: 'And being one, it is in every way alike; for if it were unlike, then, being more, it would no longer be one, but many.' Similar reasoning appears in B7.2 (what is cannot perish, grow, be rearranged, or suffer pain or grief) and the controversial B9 (what is cannot have parts). The idea that it is problematic for the same thing to be both one and many may go back to Zeno: see Eudemus *apud* Simpl. *In Phys.* 97.11–21 (= fr. 37a Wehrli; partially repeated at 138.31–139.3).

[47] Cf. also Plato, *Soph.* 251a5–c6, where the Eleatic Visitor discusses the Late-learners' enthusiastic encounters with the same problem. Note, however, that the Late-learners should be distinguished from the 'more recent' thinkers of *Phys.* I.2. The problem is the same, but the response is different. The Late-learners 'delight in not letting us call a human good, but <only> the good good, and the human human' (251b9–c2).

[48] For this problem see also Plato, *Parm.* 129c4–d6, and *Phil.* 14d4–e4.

Both of these problems can be addressed by distinguishing different ways of being one and many. For example, the second problem – the whole–parts problem – might be addressed by appealing to the distinction between being one (or many) *actually* and being one (or many) *potentially* (186a1–3). In general, a thing can be actually *F* and potentially un-*F* at the same time. A cup of tea may simultaneously be actually hot and potentially cold. Similarly with regard to one and many: a thing may simultaneously be actually one and potentially many. Socrates, for instance, is actually one in so far as he is a single human being, but he is also potentially many in so far as he has the potential to be decomposed into his various parts. Since these ways of being one and many are not opposed to one another, there is no difficulty in our saying that Socrates has both of these properties at once.

Why does Aristotle include this discussion of these problems of one and many? Certainly the discussion is prompted by his examination of the Eleatics – the worry about being one and many at the same time is Eleatic in origin. Nevertheless, the present passage does not seem to contribute directly to his refutation of Eleatic monism. Instead, it seems more likely that he is here thinking ahead to the theory of principles that he will argue for later in *Physics* I. That theory presupposes that it is possible for something to be both one and many at the same time. The underlying substratum is one in number but many in form or account,[49] and the generated natural substance is one in so far as it is a single unified object, but also many in so far as it is a composite of the substratum and the form. By taking the opportunity to address the persistent concern that being one is incompatible with being many, Aristotle is thus removing a potential obstacle to his own theory of principles.

[49] See *Phys*. I.7.190a14–17 and 190b23–4.

Physics I.3

Laura M. Castelli

I. Preliminary Remarks

Physics I.3 presents a series of objections against monistic arguments in the wake of the discussion introduced in I.2. Both I.2 and I.3, despite their differences, are ruled by the same methodological proviso introduced at the beginning of I.2 (184b25–185a20): Parmenides' and Melissus' claims, while somehow relevant to physical investigations, are not discussed as parts of a physical theory, but rather by resorting to the same tools that are used to investigate any thesis put forth 'for the sake of the argument' (*Phys.* I.2.185a5–8). This strategy is explicitly announced both at I.2.185a9–12 and at I.3.186a7–10, which are close to identical.[1] I.3 clearly follows this strategy, while there is an issue as to whether the same applies to I.2 and, if so, as to how I.2 and I.3 differ.[2] I shall address this more general issue in section VII after an analysis of the arguments in I.3.

II. Against Melissus

Greek commentators give a detailed reconstruction of Melissus' arguments in the form of standard syllogisms[3] and provide a correspondingly technical discussion of their false premises and fallacies. Although these materials are of great interest and at least Simplicius seems to report directly from

I am grateful to Andreas Anagnostopoulos, István Bodnár, Philipp Brüllmann, Tim Clarke, Lindsay Judson, Christopher Noble and Christian Pfeiffer for commenting on previous versions of this chapter.

[1] There are three differences: first, the text in I.3 adds the subject 'their arguments' at 186a8, which in I.2 can be supplied based on the immediately preceding sentence (whereas Parmenides and Melissus are the subject of the immediately preceding sentence in I.3); furthermore, I.3.186a10 has *talla* and *outhen* instead of *ta alla* and *ouden* in I.2.185a11–12. Editors' views differ on whether the first or the second occurrence of these lines should be secluded. Bekker eliminates the lines in I.2, while Ross argues in favour of keeping the lines in I.2 and secluding them in I.3.

[2] Greek commentators distinguish between I.2, which in their views discusses the opinion (*doxa*) of the monists, and I.3., which discusses the way in which it is argued for (*kataskeue*).

[3] Philop. *In Phys.* 50.29–52.11; Simpl. *In Phys.* 103.7–104.27.

Melissus' book,[4] Aristotle's text is much less explicit and it is difficult to get a clear idea of the precise formulation (if any) of the arguments Aristotle had in mind. Aristotle's comment about Melissus' arguments, that if one concedes one absurdity, the rest follows (I.2.185a11–12, I.3.186a9–10), may suggest the idea of a chain of arguments. Similarly, the formulation of the list of objections at 186a10–22 (see a13: *eita*; a16: *epeita*; a18: *epeita*) may suggest that Aristotle is following the structure of such a chain.[5] All objections fall under two types: logical mistakes (fallacious inferences) and false premises.

(i) a10–13: Aristotle imputes to Melissus the fallacy of the consequent (cf. *SE* 167b13–20, 168b35–40, 181a27–30). The mistake consists in assuming that, if S follows from P, then not S follows from not P. Melissus would infer (according to Aristotle)[6] that what does not come to be does not have a principle from the claim that what comes to be has a principle.

(ii) a13–16: Aristotle's criticism can be taken as an objection to the universal claim that there is a principle (or a beginning) of everything that comes to be.[7] This is a premise ('what comes to be has a principle') of the argument under attack in (i). 'Principle' (*arche*) is supposed to indicate a part from which change starts, rather than a beginning in time.[8] In particular, the argument seems to be about a part that changes before the rest of the subject of change.[9] The universal claim can be shown to be false through counter-examples: some changes occur all at once in the whole object, without any part's changing before the rest (cf. *Phys.* VIII.3.253b23–6).

(iii) a16–18: The conditional formulation ('why should it be immobile, *if* it is one?') suggests that the objection pertains to a relation between propositions. However, there is a general difficulty in (iii) and (iv) of telling

[4] Simpl. *In Phys.* 111.18ff.

[5] Horstschäfer 1998: 89–90 particularly insists on the characterisation of Melissus' arguments as an *Argumentationskette*, ending with the conclusion that being is unchangeable (or that change is impossible).

[6] Simpl. *In Phys.* 105.24ff. claims that Melissus did not infer the claim that what does not come to be does not have a principle from the claim that what comes to be has a principle, but had independent reasons to assume the first claim.

[7] Commentators agree that this, i.e. 'there is a principle of everything *that comes to be*' (rather than: 'there is a principle of everything') is the claim Aristotle ascribes to Melissus. On this point see in part. Ross 1936: 471.

[8] Cf. Philop. *In Phys.* 59.26–60.17; Simpl. *In Phys.* 105.32–107.38.

[9] And not about a constituent which may be a principle of change without undergoing change itself (e.g. the *physis* of natural beings is an immanent principle of change and rest, but is not a part that changes).

whether Aristotle's objection is that Melissus is assuming a false premise or that he is making a logical mistake. For it is not clear whether Aristotle is objecting to a claim such as 'alteration does not belong to what is one' or to an argument such as 'What is is one; therefore, what is does not undergo alteration'.[10] Either way, the objection is that there are cases of beings that are one (in the sense of being qualitatively/formally homogeneous) and yet are in motion (e.g. a vortex of fluid).

(iv) a18–22: That alteration is impossible does not follow from the premise that what is is one. Melissus claims that what is one (again: in the sense of being qualitatively/formally homogeneous) cannot undergo alteration. Aristotle argues that qualitative/formal homogeneity can belong to things in two ways: either (iv.a) in that their substrate is qualitatively homogeneous (while things made of that substrate display some qualitative/formal difference), or (iv.b) in that they lack any qualitative/formal distinction whatsoever. If Melissus' claim is understood as in (iv.a), then the argument is a non sequitur: in fact, several philosophers of nature assume the existence of a qualitatively homogeneous substrate (be it water or air or fire) and still make room for alteration as a qualitative modification of such a substrate. But in order to deny the possibility of alteration Melissus must have something else in mind – presumably something like (iv.b). However, if the claim is understood as in (iv.b), then the argument is sound, but it does not apply to what is, since it is evident that what is displays qualitative/formal plurality. If Melissus denied the latter, he would end up supporting a Heraclitean position (cf. *Phys.* I.2.185b19–25): but – Aristotle objects – it is evident that there is qualitative/formal difference between, for example, horses and human beings, or between opposites.[11]

(i) and (ii) focus on the starting point of Melissus' argument, i.e. the claim that everything that comes to be has a principle: (i) shows that it does not entail that what does not come to be does not have a principle, and (ii) shows that it is not true. (iii) and (iv) rather focus on the further inference that what is cannot change if what is one. While the logical mistake that Aristotle detects in (i) is merely formal and does not depend on any specific conception of being in general or of natural being in particular, it is less clear that the same consideration applies to (ii)–(iv).

[10] Melissus B7 (Simpl. *In Phys.* 111.18ff.) seems to support the second formulation. Simplicius reports Melissus' argument by saying that, if being were such as to undergo alteration, it would not be one.

[11] My reconstruction of a18–22 as a unified argument differs from that of most commentators, who tend to divide the text into two relatively independent arguments, the first at a18 and the second at a19–22. It does not seem to me that a19–22 introduces a new argument.

However, the observations Aristotle relies upon in (ii)–(iv) are taken as pretty much undisputable data: there are cases of objects changing all at once; we can observe vortices of homogeneous fluids; there exist formally/qualitative different objects. Furthermore, other philosophers before Aristotle have already acknowledged that a kind of unity of the substrate is not incompatible with change.[12]

In spelling out (iii) and (iv) I have suggested that the unity of being which figures in the premises of the arguments denying the possibility of change be understood as formal/qualitative homogeneity. *Metaph.* Α 5.986b18–20 states that, while Parmenides got closer to the notion of what is one in account (see below sections III.1, IV.1–2), Melissus got closer to the notion of what is one according to matter. Unity of matter is linked to two ways of being one in Aristotle. Firstly, in *Metaph.* Δ 6.1016a17–24 unity in substrate is described as a sort of qualitative unity of matter which is undifferentiated (or indistinguishable) to the senses. Secondly, unity of matter is sometimes used to characterise the numerical unity of the individual (*Metaph.* Δ 6.1016b31–5; cf. *Metaph.* Δ 9.1018a6–19, Λ 5.1071a27–9, Ι 3.1054a34–5). I take it that both notions of unity of matter may be in play when Aristotle says that Melissus got closer to the notion of what is one in matter: for both the unity of what is as something homogeneous (*homoion*) and the unity of what is as something full or continuous (i.e. admitting of no void) and, in this way, numerically one, seem characteristic of Melissus.[13] Note that unity understood as lack of formal/qualitative differentiation of a substrate is closer to unity in account (i.e. Parmenides' notion of unity of being) in that both homogeneity of substrate and unity in account are (also) forms of unity in *eidos*. However, the arguments against Parmenides in the second part of the chapter suggest that unity in account understood as formal homogeneity is not enough for Parmenides' monism: for in order to get Parmenides' monism one should also eliminate the plurality in account which is imported by the distinction between subject and property. I do not think this implies that plurality in account in this sense is irrelevant for the criticism of Melissus' monism; rather, Aristotle seems to think that Melissus' arguments can be discarded on the basis of less sophisticated objections (see, for example, the form in which plurality in account figures in (iv.b) above).

[12] Cf. *Metaph.* Α 3.983b18–984a8, 984a27–b1, Α 8.988b22–3; *Phys.* I.3.187a10–I.4.187a23.

[13] For an overview of the evidence see Horstschäfer 1998: 86–7.

III. Arguing against Parmenides

III.1 Structure of the Argument and Initial Criticism at 186a22–32

In approaching Parmenides' arguments Aristotle makes two claims: first, the argument concluding that all things (or things that are) are one requires the assumption of the false premise that being is said without qualification (186a24–5); second, one can show that the desired conclusion does not follow from Parmenides' premises (186a25–32).

Where does Aristotle question the premise that being is said without qualification? Perhaps in I.2? Strictly speaking in I.2 Aristotle takes issue with the claim that 'all things are one' (185a22: *hen ta panta*) and 'the universe is one' (185b7: *hen to pan*), which I take as equivalent to 'things that are are one' (I.3.186a5: *ta onta hen*). There are reasons for taking these formulations as the conclusion (rather than as equivalent to one of the premises) of the argument Aristotle discusses in I.3 (see below). More generally, it does not seem to me that in I.3 Aristotle primarily tries to show that it is false, or why it is false, that being is said without qualification. Rather, I shall argue that his discussion of Parmenides' arguments only makes sense if Aristotle distinguishes between two prima facie equivalent claims:

(a) Being signifies one thing (*hen semainei*).
(b) Being is said without qualification (*legetai haplos*).

I shall spell out what (a) and (b) mean in the following sections. By way of introduction, I take it that (a), while compatible with (b), is weaker than (b) and does not imply it.[14] In order to show that Parmenides' argument does not establish its conclusion Aristotle starts by ascribing to Parmenides an argument having (a) as one of its premises. (a), according to Aristotle, is too weak to get Parmenides' desired conclusion that things that are are one. He shows this by arguing, at 186a26–9, that the truth of the premises of the arguments is compatible with two kinds of plurality and, therefore, with the falsity of the conclusion. At 186a32–4 Aristotle then goes on to explain that what one would need in order to get something closer to Parmenides' desired conclusion is (b). But (b), according to Aristotle, is false. Furthermore, at 186b14ff. Aristotle shows that not even (b) is enough to establish a specific form of monism (borrowing Tim Clarke's terminology: (b) is not enough to establish entity monism). Note that in

[14] The argument at 186b14 also shows that (b) does not imply some specific reading of (a) either.

my reconstruction 186a34–b14 present arguments that a monist might be willing to accept. In particular, those arguments show how the monist can avoid Aristotle's initial criticism at 186a22–32. It is only at 186b14ff. that Aristotle shows that plurality cannot be excluded even if one manages to avoid the initial criticism.

Aristotle illustrates his initial strategy by sketching an argument at 186a26–7. Schematically, the propositions relevant for the formulation of the argument are the following:

(1) Only pale things are.
(2) The pale signifies one thing.
(3) Pale things are one.

Aristotle intends to show that (3) does not follow from (1) and (2). The corresponding argument for being is, therefore:

(1′) Only things that are are.
(2′) Being signifies one thing.
(3′) Therefore, things that are are one.[15]

I shall address two of the difficulties that (1)–(3) poses. First, what kind of relation is the relation of 'signifying' featuring in (2)? Second, how do (1) and (2) relate to each other? In the Greek text, (1) appears as one assumption of the argument (186a26), whereas (2) appears separately in a different grammatical construction (a genitive absolute at 186a26–7).

III.2 'Signifying' *(semainein)*

Sentences such as 'the pale signifies one thing' (*to leukon hen semainei*) or 'being signifies one thing' (*to on hen semainei*) are ambiguous in a number of ways.[16] To start with, the singular neuter adjective (*leukon*) or participle (*on*) preceded by the singular neuter determinate article (*to*) may (but need not) be used to mention the linguistic expression that follows. If one thinks that this is how such sentences are construed, one should translate: '"white" signifies one thing', or '"being" signifies one thing'. This may seem a desirable translation for a sentence about something 'signifying' or 'meaning' something. However, ancient Greek and Aristotle do not

[15] Accordingly, I take 'all things are one', 'the universe is one', 'things that are are one' as different formulations of the conclusion of the argument.

[16] Similar considerations apply to sentences such as 'being is said in many ways' or 'the one is said in many ways'.

systematically distinguish between use and mention. There are contexts in which Aristotle seems to make a point about language while he is actually making a point about reality.[17] Furthermore, *to on* or *to leukon* can pick out either a general property (being or whiteness) or the bearer of that property (what is or what is white; cf. I.3.186a28–32; *Metaph.* Z 6.1031b22–8).

The ambiguity of the subject in sentences such as (2) links to the difficulty of assigning a precise meaning to the predicate ('to signify one'). For if one takes the subject of (2) to be a linguistic expression ('pale'), then the predicate would say of the subject that it has only one meaning (i.e. that it is not ambiguous) or that it refers to one thing only. If, on the other hand, one takes the subject of (2) to be a property or an object, then the predicate would say of the subject that it can be used as a sign for one thing (see below).

Keeping these difficulties in mind, one relatively natural reading of (2) would be to take it as stating that 'pale' is univocal. There is one single property which corresponds to being pale for all things that are pale, and this is what 'pale' means. I shall label this account 'meaning-approach'.[18] The main problem with the meaning-approach is that this interpretation does not make much sense of the use of *semainein* at 186b10–12: 'What is precisely being, then, is not: for it is true to say that it is pale, and the latter signifies what is not. So that the pale signifies what is precisely being, too: therefore, being signifies several things.'

I shall return to the details of the argument in section IV. For the meaning-approach it is hard to explain how one could ever infer that the pale signifies what is not or that the pale signifies what is from the claim that the object that is pale also happens to be other than being or that the object that is pale also happens to be something that is. Of course one could say that at 186b10–12 *semainein* is used in a particularly weak and possibly idiosyncratic way, but in trying to make sense of Aristotle's use of *semainein* in I.3 I would rather not start with that assumption. My strategy will rather be the opposite: I will start from 186b10–12 to see whether a coherent account of the relation of signification can be provided which fits that passage and the rest of the chapter.

[17] One well-known case is *Metaph.* Δ, where Aristotle does not mainly distinguish the several meanings of words such as 'principle', 'cause', 'one', etc., but spells out the different ways in which things are principles, causes, one, etc.

[18] Cf. Ross 1936: 340, 473; Wagner 1967: 407. Horstschäfer 1998: 99 sees some of the problems with this approach and suggests that *semainein* covers both meaning and reference.

One further potential difficulty is that it may be tempting for the meaning-approach to take (2) as equivalent to 'the white is one in account'. If so, then Aristotle's criticism of the argument in (1)–(3) turns out to be that (2) is false. For *to leukon* for Aristotle is ambiguous in the sense that it can refer to the property or to the bearer (see 186a28–32). But Aristotle's criticism in this section seems to be that the conclusion does not follow from the premises rather than that one of the premises is false.

Finally, the meaning-approach seems to rely on the distinction between property and subject which, according to Aristotle, Parmenides did not appreciate. If the distinction is needed in order to understand the premises of the argument, then it seems necessary not to take the argument as one Aristotle could ascribe to Parmenides. On my reading, instead, the distinction between property and subject is not required to make sense of the premises of the argument and, in this sense, what Aristotle criticises could be a Parmenidean argument according to Aristotle's own views about Parmenides. I shall leave aside the issue whether the deflationary reading of the relation of signification I propose is Aristotle's standard understanding of signification. Independently of whether this is the case, in I.3 Aristotle might have in sight Parmenides' reference to 'signs' (*semata*) in fr. B8, where any attribute of being is taken as a sign of it. If so, the relatively simple account of signification provided below is at least compatible with Parmenides' use of the language of 'signs' and, therefore, is all Aristotle needs to discuss Parmenides' views.

Let us now return to 186b10–12. The development of the argument at 186b4–12 shows that Aristotle moves quite freely between saying 'it is true to say that something, x, is y' or 'x is y' and saying 'y signifies x'. For example, at b8–9: 'the pale is not being' and at b11: 'the pale signifies what is not'; or at b10: 'it is true to say that <what is> is pale' and at b11–12: 'the pale signifies what is precisely being'. This suggests that Aristotle is operating with a rather simple relation of reference between an object[19] (what is referred to) and any expression indicating a feature of it, or a relation of (for lack of a better word) indication between an object and any feature of it that can be used to flag it. If we take the relation of signification as a relation of reference, the linguistic expression signifying the object can be regarded as a name for it (see below). If we take the relation of signification as a relation of indication, the feature signifying the object can

[19] By 'object' I simply mean a chunk of reality that can be picked out as distinct from its surroundings or from the background.

be regarded as a sign for it.[20] In what follows, I shall simply speak of a relation of signification. Anything that can appear in place of y in 'y signifies x' or in 'x is y' can be thought of as a 'sign' (or an indicator or a signifier)[21] of the signified object. I take it that all the relevant features of such a relation apply to both interpretations in terms of reference or indication and that some oscillation between the two interpretations may occur in the course of I.3.

The relation of signification so understood has a few important features. First, it does not distinguish between names and predicates as two different classes of signifiers. Accordingly, both proper names and descriptions of an object x can be used as 'names' of it. This picture squares with the idea we occasionally find both in Plato and Aristotle that one and the same thing can have many 'names' (*onomata*), where the latter include any qualitative description of the object in question.[22]

Secondly, the relation of signification does not display the plurality of different ontological relations that may obtain between the indicated object and the various features picked out as signs for it. Propositions of the form 'S is P' can either express a relation of sameness or a predicative relation. Broadly speaking, 'x is y' is true if x and y are at least partially overlapping chunks of reality. Aristotle distinguishes between different predicative relations and different ways to convey numerical sameness, and this implies that for Aristotle 'x is y' can be true under very different conditions. For example, consider the case in which 'x is y' is a predicative proposition in which x indicates the subject and y the predicate. Since predicates can stand in different relations to their subjects,[23] y can indicate an essential feature, a coextensive feature or an accident of x. To all these predicative relations there correspond ways to convey numerical sameness,[24] and different pairs of expressions can be used to formulate claims of numerical sameness.[25]

[20] The necessity to spell out these two different relations is a consequence of the lack of a neat distinction between use and mention in Aristotle.

[21] Note that the literal meaning of the verb *semaino* is 'to show by a sign, indicate, point out'; the noun with the same root *semeion* means 'mark' or 'sign' or 'trace' from which something is known.

[22] *Soph.* 244b9ff., 251a5–c6, 255a1–2, c9–11; *Phil.* 14c7ff.; *Parm.* 128e6–130a2. Cf. Aristotle's description of numerical sameness in *Top.* I.7.103a9–10.

[23] For the distinction between kinds of predicate, see *Top.* I.5 and 8.

[24] For the different ways to convey numerical sameness see *Top.* I.7.103a23–39. The pair subject–genus is not discussed in *Top.* I.7 as conveying numerical sameness; however, Aristotle's use of *hoper* (see references in section IV.1) to indicate the relation between a subject and its genus suggests that he envisages a sort of 'vertical' identity between a subject and all the predicates that express its essence. I am indebted to Mauro Mariani for discussion of this point.

[25] Note that, if my analysis of signification is correct, it does not make a big difference whether we retain the text with *tauto* at 186b6.

Thirdly, in order to express verbally a relation of signification between a determinate object and any of its signs the object has to be picked out by one of them (setting aside the case in which one can point at the object and say '*This* is white' or 'the white signifies *this*'). Any sign can be used for this purpose. From this it follows that, for any pair of signs P and R of an object x (or: for any pair of expressions P and R such that it is true to say that x is P and that x is R), we can say that P signifies R as well as that R signifies P, provided that the expression on the left of 'signifies' is taken as picking out a feature (or mentioning a name) of x, while the expression on the right of 'signifies' picks out the object x. For example, if it is true of something x that it is white and that it is one, we can say all of the following:

(a) 'Pale' signifies x. / The pale signifies x.
(b) 'One' signifies x. / The one signifies x.
(c) 'Pale' signifies what is one. / The pale signifies what is one.
(d) 'One' signifies what is pale. / The one signifies what is pale.

All of these features of signification will be important if we are to appreciate various aspects of Aristotle's arguments at 186a34–b12.

One further relevant passage is 186a32–4. Aristotle's remarks make it clear that, in order to say that P signifies one thing, it is enough if (i) it is true that P is predicated of something (possibly by being one of its accidents) and (ii) it is also true that that thing, of which P is predicated, is one. This leaves room for an ontological distinction between the thing that is P and P itself (see section III.3 below). Aristotle says that this is a distinction Parmenides did not see. This is why Aristotle then finds an interpretation of Parmenides' premise that being signifies one thing which does not leave room for such a distinction. This can be done if the claim is understood in such a way that neither being nor unity is an incidental property of something which is and which is one. Rather, 'being' and 'one' have to signify being itself and unity itself. The way in which Aristotle introduces this understanding of Parmenides' claim (see section IV) strongly suggests that this is *not* the only or the standard interpretation of propositions such as 'being signifies one thing'. From the development of the argument we can also infer (sections IV.1 and IV.2) that this understanding of Parmenides' claim corresponds, for Aristotle, to saying that being as such is (precisely) a substance.

III.3 The Relation between (1) and (2)

Let us now take a closer look at (1), its relation to (2) and the reasons why, according to Aristotle, (3) would not follow from (1) and (2). Claim (1) as such does not appear in the text, but it seems to be what is meant at 186a26, literally: 'even if one assumed pale things only'. I take (1) in the existential sense:

(1*) Only what is pale exists.

Aristotle suggests that Parmenides' understanding of (1) is compatible with the further claim:

(1a) There is nothing beyond pale things.

(1a) is taken in the sense that there is nothing separate from pale things. In particular, 186a28–32 and 187a7 suggest that for Parmenides (1a) implies (1b):

(1b) There is nothing apart from pale things which can be said to be pale.

In order to make sense of Aristotle's reconstruction of Parmenides' argument here, we can imagine a completely white continuous world, i.e. a world displaying no gaps in whiteness[26] and in which no contiguous objects exist. In this world, no plurality of separately identifiable referents for 'white' can be found.[27] With respect to this world, (2) says that the object that is white is also one. The referent of 'white' is one because there are no separately identifiable referents that can be counted (for how could they be distinguished from each other in the homogeneous white and continuous landscape?).

Aristotle makes two objections, showing that the truth of the premises of the argument is compatible with some form of plurality. First, even if there is only one identifiable white object (i.e. the white world) which is signified by the white, that object can be spatially divided. The thought behind this objection is, presumably, that colour is a per se property of surfaces, which are a kind of magnitude. If something is white, then it has magnitude. Anything that has magnitude can be divided[28] and, by being divided, it will turn out to be many.[29] Secondly, (1a) does not imply (1b): it is possible to distinguish the thing that is white from its being white. The fact that the property of being one does not work as an independently

[26] 'Pale' is, in general, a better translation for *leukon*, but for the sake of simplicity I construe the example in terms of whiteness, rather than in terms of paleness, because paleness, for modern readers, is compatible with difference in hues, whereas Aristotle associates differences in hues with differences in brightness.

[27] Note that this world structurally resembles Melissus' homogeneous and full (i.e. continuous) being.

[28] On divisibility as a distinctive feature of quantity see *Metaph.* Δ 13.1020a7–8.

[29] Note, however, that the existence of a single continuous (i.e. non-divided) white object complies with the conditions under which it would be true that the white signifies one thing as spelled out above.

identifiable referent for 'white' does not imply that there is nothing white apart from white things: the colour white is called 'white' and is ontologically distinct from the things that are white.

IV. Avoiding Aristotle's Initial Criticism and What is Precisely Being (186a32–b13)

Having shown why Parmenides' argument does not establish the desired conclusion, Aristotle shows that a stronger assumption than (2) is needed (186a32–4) in order to avoid the objections mentioned at 186a26–9. The stronger premise must be such that no room is left for plurality in being between a subject that is and the property of being (186a34–b12), and that the possibility of conceiving being as something continuous and, therefore, divisible is also excluded (186b12–14). If this is right, then the arguments down to 186b14 are arguments the monists would (or should) accept.

IV.1 *to hoper on* (186a32–4)

In order to avoid the objections mentioned at 186a26–9, one should not only assume that being signifies something that also happens to be one (for this leaves room for forms of ontological plurality). Rather, the object picked out by 'being' must be precisely being (*hoper on*) and precisely one (*hoper hen*). This means that what 'being' and 'one' signify must be such that being and being one fully express its nature.[30]

If the account of signification I have suggested above is correct, then the claim 'being signifies one' says no more than the object(s) of which being is predicated also happen(s) to be one. This leaves room for two possibilities which should be excluded: 'being' and 'one' (or being and one) could refer to (or indicate) accidents of the object they signify. This would imply that the thing that is and the property of being that belongs to it are different. Furthermore, the relation between being and being one could be a merely incidental one. Consider the sentence:

'The white signifies the just'.

This claim is true if there is something of which it is true to say that it is white and of which it is true to say that it is just. Both being white and

[30] This result, i.e. that being and being one fully express the nature of being, is not granted by the use of *hoper* alone: the human being is *hoper* animal, but this does not imply that being an animal exhausts and fully expresses the nature of the human being. Rather, that 'being' (or being) must signify what is precisely being is the conclusion of the arguments presented in this section.

being just could be merely incidental properties of their subject and the relation between being white and being just could be an incidental relation.[31] Any incidental relation between A and B presupposes that A and B are different (and, therefore, many) in being.

Aristotle writes that 'it is necessary to assume not only that being signifies one thing with reference to that of which it is predicated, but also that <it signifies> what is precisely being (*to hoper on*) and what is precisely one (*to hoper hen*)' (186a32–4). The formulation *hoper* + a noun or an adjective is often used in the *Topics* to express essential predication: if S is *hoper* P, then the definition of P applies to S. In particular, *hoper* is used to introduce the predicate in sentences where the predicate is a genus of the subject (*Top.* 116a23ff., 120b23, 122b25–26, 128a35, 141a35–7, 146b4, 147a14). Since antiquity there has been a debate among commentators[32] about whether such a use of *hoper* is really relevant to I.3 or one should take the formula *to hoper on* as an unusual formulation to express what is properly (*kurios*) and mostly (*malista*) being. In this sense the expression would indicate being itself (*to auto on*) for the Platonists and substance (*ousia*) for Aristotle.[33]

It does not seem to me that the two options are incompatible: substances for Aristotle are objects which do not differ from what they are (or their essence). Aristotle sometimes expresses this feature of substances by saying that substances are without being something else. We have seen that, according to Aristotle, Parmenides has to avoid a conception of being that leaves room for a distinction between the property of being and the thing which has that property. This can be achieved if one conceives of being as something whose essence consists precisely in being. The fact that for Aristotle objects of this sort are substances is made particularly clear by the example he uses at 186b14ff., but the identity between a subject and its essence is what really matters in the formulation at 186a32–4.

IV.2 Formal Features of What is Precisely Being (186a34–b12)

What is precisely being enjoys two formal features: it cannot belong to any subject which is other than what is precisely being[34] (186a34–b4); and no accidents can belong to it (186b4–12).

[31] For the two cases (incidental relation between two attributes and incidental relation between one attribute and its subject) see *Metaph.* Δ 6.1015b17–34, Δ 7.1017a8–22, Δ 9.1017b27–33.

[32] See, for example, Simpl. *In Phys.* 122.25ff. and, in part., 123.16ff.

[33] Simpl. *In Phys.* 122.25–32.

[34] Note that Aristotle ascribes to Parmenides the failure to distinguish between otherness or difference and separation: if such a distinction is not in place, the fact that there is nothing separate from what

The first point is argued for in the following way. Suppose that being is an accident of some object x; x will be different from being and, therefore, it will not be being. At the same time, if it is true that being is an accident of x, it will be true to say that x is (or that x is being) and, therefore, that being signifies x in addition to signifying what is precisely being. Therefore, being will signify more than one thing. The most difficult line is 186b2, which I read in the following way: 'For it will not be possible to be it [sc.: what is precisely being] while being something determinate [sc.: by being something independently identifiable], unless being signifies several things in such a way that each of them is something,' i.e. unless the premises that being signifies one thing only and, in particular, that being signifies what is precisely being are dropped.

As for the second point, we have seen above that the relation of signification is such that, if it is true to say of something x that it is y, then y signifies x. Let us assume that being has the property of being pale, i.e. that being is pale. One can argue that the pale is not being, since neither being pale is the same as being being nor can being be incidentally predicated of the pale (for the reasons spelled out at 186a34–b4). One can then infer the claim that the pale signifies what is not being (b11) from the claim that it is true to say that the pale is not being.

At the same time, since it is true to say, *ex hypothesi*, that being is pale and the pale signifies what is not being, it will follow that, on the one hand, what is precisely being is what is not being (b10) and, on the other hand, that the pale (i.e. what is not being) will signify being. Given that being is other than what is not being, being will signify more than one thing, i.e. what is precisely being and what is not being. The reason why being will signify what is not being is not made explicit in the text, but if the account of signification above is correct, this is because it is true to say of the thing which is pale (i.e. of the thing which is what is not being) that it is also being.

This pair of arguments shows that the only conception of being which is compatible with Parmenides' lack of distinction between subject and property is one where being is just and precisely being, without being anything else. This conception of being is compatible with the claim that being signifies one thing but actually requires more than such a basic claim: it requires that being be precisely this, i.e. being. In Aristotle's ontology, the objects that get the closest to lacking any ontological

is precisely being collapses into the claim that there is nothing other than what is precisely being. See comments on 186a29–32.

structure introducing a distinction between a subject and a property are substances. In this sense, what Parmenides' argument requires is a conception of being according to which being is understood absolutely, i.e. as a simple substance. However, the fact that what is precisely being can only be understood as a simple substance does not play any role in the argument up to this point; rather, it will become evident in Aristotle's choice of examples in section 186b14–35.

I take it that both arguments would be acceptable for true monists. In particular, I take it that the second argument shows that nothing can be said of the one in that any relation of predication introduces plurality. In this perspective, the monists are supposed to grant the premise that something (the pale) is predicated of or is an accident of being only as the beginning of a *reductio ad absurdum* which they endorse.[35]

IV.3 The Lack of Magnitude of Being (186b12–14)

Ancient commentators take b12–14 as the premise of a further argument, left implicit in our text: if being has no magnitude, it cannot be finite (contrary to what Parmenides claims) nor can it be infinite (contrary to what Melissus claims). On this reading, the lack of magnitude of being is not a result the monists want. Rather, if the monists assume the stronger premise about the absolute nature of being which is necessary to avoid the initial criticism, the unwelcome conclusion will also follow, that being will not have magnitude.

I suggest that b12–14 complies with the strategy of the whole section starting at 186a32: Aristotle's understanding of what is precisely being is what is needed in order to avoid the criticism at 186a28–9. In this perspective, if the lack of magnitude is a feature of being that the monists want (or should want),[36] Aristotle shows that this follows from the stronger premise that being is what is precisely being (but does not follow

[35] Whereas if one takes the argument as a *reductio ad absurdum* of monistic positions one should explain why the monist should grant that something is an accident of being. This problem is stressed by Bostock 2006b: 111, who, however, follows the traditional reading of the argument, according to which the argument sets out a problem for the monists. But Bostock 2006b: 110 sees that the argument down to b4 'seems to deliver the result that Parmenides wants'. I take it that the whole discussion down to b14 is supposed to deliver results that the monists want or should want.

[36] Whether this is the case is controversial; see *Phys.* I.2.185b16–19. In fact, Aristotle's arguments in *Phys.* I.2 strongly suggest that the monists take being to be spatially extended. It is possible that I.3 exploits a gap between what genuine monists should claim and what Eleatic philosophers actually claimed. I am grateful to Tim Clarke for discussion on this point.

from the weaker premise that being signifies one thing). The problem with the stronger premise, is that, according to Aristotle, it is false.

There is an issue as to whether the two parts which are different in being are being and its magnitude, if it had any, or the quantitative parts.[37] *toinun* at b12 suggests that the passage expresses a consequence of what has been shown in the preceding lines. What has been shown is that what is precisely being is not compatible with any distinction between subject and property. Accordingly, the two parts which would be different in being, if being had magnitude, would be being (the subject of magnitude) and magnitude (an incidental property of being). From the fact that being cannot have magnitude it also follows, however, that being cannot be divided into quantitative parts.

V. The Division of What is Precisely Being into What is Precisely Some Being (186b14–35)

This section is usually taken as an argument that Aristotle uses to show that being turns out to be many even if it is not understood as a compound of subject and property. Assuming that being is what is precisely being, it can still be divided into parts, each of which is what is precisely some being, i.e. into parts of being which do not introduce a distinction between subject and property. Such a division of being 'per se' into such parts is exemplified by the division of the definition of a substance into its parts: no part of the definition of the substance is an accident of something else. On this reading, the argument is supposed to establish two results: first, it confirms what has already been shown in the former lines, namely that what is precisely being cannot be a compound of two items, a subject and an attribute, which are different in being. Second, it shows that, even if one eliminates this kind of plurality, it is still possible to divide being into parts, and this division, while compatible with the assumption that being is what is precisely being, is incompatible with the monistic denial of any plurality in being. The main problem with this reading is that it does not explain how 186b35 ('the all is therefore made of indivisibles') is supposed to follow from the argument.[38] Furthermore, the transition to this section from the former section is not particularly clear. If the point of the

[37] Cf. n. 28 on quantity and divisibility.

[38] Ross 1936: 478–9 suggests that this need not be a statement. Rather it can be taken as a question. In this way, he says, one can avoid the problem of explaining how the statement follows as a conclusion from what precedes. However, even if one accepts Ross's move, one still wants to be able to track the train of thought leading to the formulation of the question.

argument had been to show that what is precisely being can be divided and is, in this sense, many, this could have been made explicit. What we find instead is the explicit insistence on the claim that being can only be divided into what is precisely being and a puzzling inference from this claim to the alleged conclusion that being should be divided into indivisible parts. I shall try to show that, while we can retain the relatively uncontroversial reconstruction of the structure of the argument, in order to make better sense of its location in the chapter and of the puzzling conclusion at 186a35, a slight shift of emphasis is needed.

The basic structure of the argument is the following. The claim Aristotle intends to establish is:

(1) What is precisely being divides into what is precisely some being.

In order to argue for this claim, Aristotle introduces an example. Take something that is precisely being, i.e. a substance: say, human being. It can be shown that:

(1a) The parts of the human being (biped and animal), too, will be what is precisely being.

The proof is a *reductio ad absurdum*:

(2) If the parts of the human being were not what is precisely being, then they should be accidents.
(3) If they were accidents, either (3a) they would be accidents of the human being or (3b) they would be accidents of something else.

Aristotle argues separately against (3a) and (3b).

The argument against (3a): They cannot be accidents of the human being because if they were:

(4) There are only two ways in which something can be an accident: (a) by being that which can belong or not belong to its subject (it can be separated from it), e.g. sitting; (b) by being that which includes the account of its subject in its definition.
(5) The parts into which the human being can be divided (= the parts of its definition, e.g. animal and biped) are not accidents in sense (b).
(6) Therefore, the parts of the human being are accidents in sense (a), i.e. the parts of the human being can be separated from it.
(7) If (6), then it is possible that the human being be not biped or animal.
(8) But (7) is impossible.
(9) Therefore, (3a) is impossible.

Against (3b):

(10) If the parts of the human being were accidents of something else, then also the human being as a whole would be something else.
(11) But (10) is impossible, because the human being is nothing else (i.e. is what is precisely being/is a being per se/is a substance).
(12) Therefore, (3b) is impossible.[39]
(13) Given that both (3a) and (3b) are impossible, it is impossible that the parts of the human being are accidents.

From the particular argument concerning the parts of the human being Aristotle infers the general conclusion that what is precisely being can only be divided into what is precisely being, i.e. (1) above.

What could this argument possibly have to do with the idea that being is made of indivisible parts?[40] The argument, as we have seen, establishes that, if what is precisely being can be divided at all, then it cannot be divided into parts performing different ontological roles, such as a subject and a property. Rather, if the division is at all possible, such a division can only be a division into parts that share with the whole its ontological status, i.e. that of being what is precisely being.

This train of thought is reminiscent of some of the tenets shared, with opposite results, by Melissus and the Atomists: Melissus (B8) suggests that, if being were many, then it should be infinitely many and each part should share the same ontological features of being as such. This, for Melissus, is absurd and provides an argument to deny plurality. The Atomists, on the contrary, just introduce an infinite plurality of indivisible beings, i.e. the atoms, each of which shares basic ontological features with the Eleatic being.

Some passages in *De generatione et corruptione* I can help us further in figuring out what the argument about the divisibility of being into being might have to do with the introduction of atomic parts of being.[41] *GC* I.2.316a14–b18 provides a reconstruction of the difficulty that led the Atomists to introduce the atoms. If one assumes that magnitude is divisible

[39] My reading of (10)–(12) is in agreement with that of most commentators. For a rather different interpretation see Horstschäfer 1998: 114–18.

[40] There is a debate as to what the indivisible parts Aristotle refers to here are supposed to be: is he talking about points? Is he talking about indivisible substances of some sort? Or is he thinking of indivisible parts of a definition? I hope to show in what follows that the exact understanding of the 'indivisibles' at stake is less important than the overall train of thought leading to the introduction of this option.

[41] On the different use of the same arguments by the monists and the Atomists see Schramm 1962: 245–64; Rashed 2005: 103–4.

throughout and that the complete division of such a magnitude has been carried out, what will remain at the end of this process? It cannot be magnitudes, since, by definition, they would be further divisible; but it was assumed that the complete division has already been carried out. But what remains cannot be points (i.e. entities without magnitude) either, nor nothing at all.[42] The detail to which I would like to draw attention is the way in which a result of the division is described at 316b2–5: 'If, then, what is detached is not a body but a certain separate form (*eidos ti choriston*) or an affection (*pathos*), and the magnitude is points or contacts affected in such a way (*stigmai e haphai todi pathousai*), it is absurd that magnitude is made of things that are not magnitudes.'

Other difficulties concerning the analysis of a magnitude resulting in the isolation of an affection are raised at 316b12–14. The discussion of the analysis of a magnitude into points and an affection of points, and its exclusion as a viable option as an alternative to the division of the all into nothing, eventually lead to the denial of the complete divisibility of magnitude at 316b14–16: 'So that, if it is true that it is impossible that magnitudes be out of contacts or points, it is necessary that there are indivisible bodies and magnitudes.'

Based on a similar argument, with the addition that the introduction of indivisible magnitude is unjustified and not viable (cf. *GC* I.8.325a8–12), the monists deny the divisibility of being altogether. We could read *Phys.* I.3.186b14–35 in light of these arguments: the monists (and all those who do not refute their arguments) might have a point when they say that, if being (i.e. what is precisely being) has to be divided at all, it can only be divided into parts that have exactly the same ontological features as being. But if this is true, and if there are independent reasons for making room for plurality through division, then it looks as if the only viable option is the atomistic option. For if the division of magnitude into points (which can be taken as affections of magnitude) or into points and some further affection (i.e. into a subject with a certain property) is not compatible with the nature of what is precisely being, and if division still has to be endorsed for independent reasons, then the introduction of indivisible beings might look like the only alternative to monism.

On this interpretation, 186b14–35 is the Aristotelian version of a part of the argument which led the monists to deny the possibility of the division of being and other philosophers to introduce atoms. If so,

[42] An extended object cannot be obtained from the addition of point to point (i.e. of something without extension to something without extension), nor from nothing.

the conclusion at b35 is less puzzling than is usually assumed.[43] Furthermore, by introducing the possibility of a division of being into beings of the same sort through the division of a substance into its formal parts and by linking this procedure to the introduction of atoms, Aristotle indirectly follows a train of thought characteristic of his predecessors. *Metaph.* N 2.1089a31–5 and 1089b15–32 underline that Aristotle's predecessors were particularly preoccupied by the possibility of a plurality of substances, i.e. by the possibility of making room for a division of being itself into a plurality of beings each of which would have the properties of being itself. In taking this approach, Aristotle's predecessors basically ignore the issue of explaining the plurality of kinds of being (substances, qualities, quantities, relatives etc.). If being is understood as what is precisely being, we can see why the only available division of being is a division into parts that share the same status as the whole.

VI. The Last Section of the Chapter (187a1–11)

At 186b35 Aristotle has introduced the idea that some specific non-monistic positions can arise by drawing some consequences of Parmenides' assumptions. In the last section of I.3 Aristotle insists on the idea that some non-monistic positions rest on the basic acceptance of some Eleatic arguments, which can be questioned without endorsing any specific view about natural being and its principles.

To start with, at 187a1–3 Aristotle underlines that 'some' have given in to two Eleatic arguments in assuming the non-Eleatic claims that non-being is and that there are atomic magnitudes. The former assumption results from the failure to see that it is not true that, if being signifies one thing, then all things are one; the latter assumption results from the failure to see the problems with some argument 'from dichotomy'. Ancient and modern commentators agree on seeing in the argument from dichotomy some argument, presumably formulated by Zeno, leading the monists to the conclusion that being cannot be divided (cf. section V). I have suggested that at least an echo of such an argument can be found at 186b14–35.

As for the identification of those who give in to Eleatic arguments, ancient commentators[44] tend to find Aristotle's opponents in Plato and

[43] Bostock 2006b: 115 also suggests that Aristotle might envisage some analogies between logical division and Zenonian dichotomy. What I propose is a way to spell out this idea.

[44] See in part Philop. *In Phys.* 84.3ff. and 87.14ff. respectively.

Xenocrates (the first would go against Parmenides in the *Sophist* by admitting the existence of non-being, while the second would introduce indivisible lines). Contemporary interpreters[45] instead think that Aristotle is actually referring to the Atomists. From *Metaph.* A 4.985b4–9 and *GC* I.8.324b35ff. and similar passages we gather that in the development of atomistic doctrines a very specific interpretation was given of the necessity of assuming non-being as a principle and of the division of being into beings. The non-being rejected by the monists is void. Since (according to the Atomists) it seems true that there cannot be change without void, the monists are naturally led to the claim that being is unchangeable. The Atomists bite the bullet: it is true that in order to account for change one has to admit the existence of void. But (the Atomists say) we have to account for change, for the existence of change is a phenomenon evident to the senses; it follows that an adequate account of nature has to make room for both being (what is full) and non-being (void). Being will then be divided by non-being into infinitely many indivisible parts, each of which will be in exactly the same way as any other indivisible part. It seems to me that both Academics and Atomists could be Aristotle's targets here. In fact, in *Metaph.* N 2.1089a1–6, Aristotle suggests that everybody who thought that a principle opposite to being is necessary in order to make room for pluralism was in some way giving in to the Eleatic way of thinking. In this sense, the refutation of Eleatic arguments is not only the starting point of a new approach to natural being and its principles: it is also the starting point of a new approach to the enquiry into being more generally.

The two remaining sections (187a3–6 and 187a6–10) make two points regarding the Eleatic argument leading to the conclusion that all things are one. These points correspond to the two concessions mentioned at 187a1–3. First, the complete exclusion of non-being does not follow from the assumption that being signifies one thing and the assumption that two contradictory propositions cannot be true at the same time. Both assumptions are compatible with the claim that there is something of which it can be truly said that it is not x in some qualified sense. This claim is enough to make room for pluralism and this implies that there is no need to try to refute monism by positing some absolute non-being.[46] Secondly,

[45] Ross 1936: 479ff.; Wagner 1967: 411; Charlton 1970: 63; Horstschäfer 1998: 118–24.

[46] Cf. *Phys.* I.2.185b34–186a3, where Aristotle comments on the apparently puzzling claims that the predecessors felt compelled to accept, not seeing that they are actually less problematic than they initially seem.

Aristotle stresses the absurdity of deriving the monistic claim that all things are one from the assumption that 'there is nothing else beyond/over and above being itself'.[47] The formulation of the assumption is interesting here. We have already seen that at 186a29–31 the construction *para* + accusative is linked to separation and diversity in being. In that passage Aristotle says that the fact that there is nothing beyond things that are does not imply that there cannot be any difference in being between things that are (even if they are not separate). The point Aristotle makes at 187a6–10 can be taken in two ways, depending on whether the use of *para* + accusative is taken to express separation. If it does express separation, then the point would be the same as at 186a29–31. Furthermore, if 'being itself' can be taken as a way to refer to substance(s), then Aristotle's point would be the following: the fact that there is nothing existing over and above substance does not imply that there are no other beings than substances. This is a point to which Aristotle himself would subscribe; this interpretation brings in some basic Aristotelian assumptions about being and its structure. If *para* + accusative does not express separation but simply ontological diversity, then Aristotle's point would be that, even if there are no other beings apart from what is precisely being (substances?), it is still possible that there is a plurality of beings. This point has been illustrated at 186b14–35.

VII. Final Remarks

To conclude, I shall briefly address two general issues. First, in I.2.185a5–8 the discussion of monistic claims is said to be relevant for the enquiry into the principles of natural being. I take it that the reason why the discussion of monistic views and arguments is relevant is that they influence the enquiry into nature[48] in a very specific sense: philosophers after the Eleatics, while rejecting monism as a whole, have not really rejected monistic premises or refuted monistic arguments. This has led them to approach natural being through a conceptual framework built on some basic Eleatic assumptions. This point is made particularly clear at the end of the chapter (187a1–11)[49] as well as in several other passages where Aristotle discusses the 'archaic' and basically Eleatic starting points of the enquiry into being of most of his predecessors and contemporaries.

[47] For a similar formulation of Parmenides' argument cf. *Metaph.* A 5.986b28–30.

[48] Aristotle thinks that they influence the enquiry into being in general, but I think that in I.3 he is primarily focusing on the development of physical theories out of monistic assumptions.

[49] Cf. *Phys.* I.2.185b25–186a3.

By challenging Eleatic assumptions and Eleatic inferences, Aristotle gets rid of a way of thinking about being, nature and its principles which goes far beyond monism. The basic Eleatic assumption shared by several other philosophers is that it is necessary to introduce a principle opposed to being – some sort of absolute non-being – which would cooperate with being in causing plurality and change. The atomists, in particular, share this assumption by introducing void as a cause of plurality and change. In doing this they reject monism by embracing a specific view about the principles of natural being. However, Aristotle seems to believe that the Eleatic premises and the logical mistakes that make monistic arguments unacceptable do not have much to do with any distinctive view about natural being as such.[50] Accordingly, the refutation of monism Aristotle proposes should be endorsed before and independently from the endorsement of any more specific views on natural being and its principles.

These considerations take us directly to the further issues of the conceptual tools deployed in I.2–3 and of the similarities and differences between the two chapters. We have seen that in I.3 Aristotle discusses the logical structure of allegedly monistic arguments[51] and the contents of the premises of such arguments. In order to give a more specific characterisation of the conceptual tools Aristotle deploys in these chapters in addition to the logical analysis of inferences, we should take a look at the basis on which he questions the premises of monistic arguments. I have suggested that, in the case of Melissus' arguments, the premises are questioned on the basis of undisputable data provided by any pre-theoretical observation of the world (cf. section II). As for Parmenides, Aristotle challenges his premises on the basis of the distinction between the bearer of a property and the property, and of some (very general) understanding of the relation of signification. These should just belong to the basic intellectual equipment of anybody ready to engage in the discussion of philosophical issues.[52]

[50] At least as far as the monistic views discussed in *Phys.* I.2–3 go. Parmenides' claims about hot and cold as causes (*Metaph.* A 3.984b1–8 and A 5.986b31–987a2) can be taken as distinctive views about the principles of natural being. Perhaps some such views are what Aristotle has in mind in *Phys.* I.3.186a22, where he refers to some 'specific' arguments which are perhaps liable to a different kind of criticism.

[51] Even if the arguments Aristotle discusses were not actually formulated by the monists, Aristotle seems to be interested in unfolding thoughts which would have encountered his interlocutors' assent if they had been explicitly formulated in front of them: cf. *Metaph.* A 10.993a22–4. In this sense, even if Aristotle is not particularly worried by issues of authenticity, he seems at least to take issues of fairness into account. Cf. *Top.* VIII.5.159b27–35.

[52] Cf. Aristotle's refutation of the detractors of the principle of contradiction in *Metaph.* Γ 4.1006a11ff.: as long as they intend to use language to signify anything, they turn out to commit to the principle they intend to deny.

One might wonder whether the same can be said about the criticism we find in I.2. In order to approach this issue we should ask a question about the theoretical status of the distinction between the different ways of being and the different ways of being one Aristotle resorts to in I.2. For one might think that such distinctions are richer, from the doctrinal point of view, than the conceptual tools for the logical analysis carried out in I.3. For reasons of space I cannot discuss this question here. However, it seems clear that, even if I.2 is doctrinally more committed than I.3, neither chapter is based on any theory of *natural* being as such. Whether the conceptual framework Aristotle resorts to in I.2 and I.3 is logical rather than metaphysical and whether there is any difference between the two chapters in this respect are questions the discussion of which I shall defer to another occasion.

Physics I.4

Cristina Cerami

Introduction

Physics I.4 is generally considered to be the chapter in which Aristotle reviews theories that had been held by Presocratic physicists about principles. After Aristotle has examined and refuted, in chapters 2 and 3, the arguments of the Eleatics, one would expect him to turn to the discussion of the opinions held by the thinkers who talked about nature 'in an appropriate way', that is, those who disagreed with the doctrine that the principle is one and motionless. According to the division given at the beginning of chapter 2 (184b15–22), this should be the second step of the inquiry into principles. In lines 184b15–22, Aristotle divides the possible opinions about principles into two groups, and associates some of his predecessors with some of these opinions:[1] (I) there is just one principle, (A) either unchangeable (i.e. the Eleatics) or (B) changeable (i.e. Ionian physicists); (II) there is a plurality of principles; (i) either limited in number (and necessarily also limited in kind); (ii) or unlimited, and in this case (α) either the same in kind and different in shape, (β) or different in kind and in shape, or even opposites.[2]

I am grateful to Andrea Falcon and Alan Code for comments and suggestions on an earlier draft. I would also like to express my gratitude to Michael Chase for revising my English.

[1] This is the way in which Philoponus, at the beginning of his commentary on I.2, reconstructs the entire 'dialectical' section of the book (at least I.2–4). He explains that Aristotle aims at providing a complete division of the possible views on principles, even if not all of them are actually endorsed (cf. Philop. *In Phys.* 20.21–5.11).

[2] In I.2 Aristotle does not mention a supplementary alternative within (II), according to which there could be a plurality of motionless principles, either finite or infinite. It is not clear whether he admits this possibility. Philoponus suggests that the view that there is a finite plurality of motionless principles is to be associated with followers of Pythagoras and of Plato, while no one advocates that there is an infinite plurality of motionless principles. As we will see, however, Aristotle seems to suggest in I.4 that assuming that there are 'many principles' also immediately implies that they are changeable.

Even if it is not entirely clear which Presocratic physicists Aristotle includes in each subdivision, one would expect him to discuss in chapter 4 the doctrines held by all of them; but this is not the case. Although at the beginning of chapter 4 Aristotle offers a similar exhaustive division of Presocratic doctrines and mentions explicitly at least three of them, i.e. Anaximander, Empedocles and Anaxagoras, the chapter is entirely devoted to the last one: Anaxagoras. Other Presocratics, and especially Democritus, are quite surprisingly absent in I.4.

In the pages to follow, I will try to explain the reason for Aristotle's choice. I shall argue that the analysis of Anaxagoras' doctrine, in the way Aristotle presents it, and its refutation are decisive for Aristotle's project in *Physics* I, not only because Anaxagoras' doctrine allows a shift from the study of beings and principles to the study of principles and change, but also for its peculiar implications. According to Aristotle, even if Anaxagoras acknowledged a certain kind of differentiation between what underlies and its determinations, his doctrine is a perfect exemplification of the inconsistency of those theories that cannot admit a proper separation between predicates and substrate. It is particularly because of this aspect that Anaxagoras' doctrine must be refuted.

The chapter is divided into three main sections: (i) at 187a12–26 Aristotle divides physicists into different groups; (ii) at 187a26–187b7 he reviews Anaxagoras' arguments in support of his own theory; (iii) at 187b7–188a18 he refutes these arguments and concludes that Empedocles' doctrine is more plausible than Anaxagoras'.

I. Aristotle's Division of Presocratic Physicists

Before taking into consideration Aristotle's account of Anaxagoras' doctrine, and in order to understand his purpose in choosing this latter as the main target of his criticism, it is necessary to examine the division given at the beginning of our chapter. This division is problematic in several respects. Actually, it is not completely clear what criteria Aristotle adopts in establishing it, and whether he picks the same criteria used in chapter 2, or instead goes beyond the results of this chapter by introducing new elements of analysis. I shall argue that the two divisions do not consider Presocratic doctrines in exactly the same way. While I.4 is clearly connected with the preceding discussion of chapters 2–3, the way in which Presocratic doctrines are considered in this chapter is quite different from one that adheres to the divisions set out in I.2. I refer the reader to my translation of lines I.4.187a11–26.

A textual discussion is in order before undertaking more general considerations. This textual point will help us in understanding the meaning of the division offered by Aristotle and its role in the general economy of book I.

In his edition, Ross corrects the transmitted text at lines a12–13:

> [a12–13] οἱ μὲν γὰρ ἓν ποιήσανες τὸ [ὂν] σῶμα τὸ ὑποκείμενον

and follows at lines a21–2 the text transmitted by Simplicius and Philoponus, the first hand of the Vindoboniensis 100 (J) and two later manuscripts (F and I):

> [a21–2] καὶ ὅσοι δ' ἓν καὶ πολλά φασιν εἶναι

Ross takes the τὸ ὂν at line a13 and the τὰ ὄντα at line a22 to be two glosses. He explains that the grammatical structure of the transmitted text at a13 is not natural and that the addition, in the same line, has been made in order to link chapter 4 to chapters 2 and 3. He also mentions the suggestion made by Neuhaüser, who takes the Greek to be equivalent to οἱ μὲν γὰρ ἓν σῶμα ποιήσανες τὸ ὂν τὸ ὑποκείμενον. Ross concludes that, because of the order of Greek words, this reading is even less plausible than the one implied by the transmitted text. Accordingly, he considers τὸ ὑποκείμενον as the subject of the two sentences and explains that Aristotle divides his predecessors' doctrines by the fact that some believe that the underlying body is one *and* that everything else derives from it by condensation and rarefaction, while the others think that the underlying body is one and many *and* that everything else derives from it by separation (ἐκκρίνεσθαι):

- The underlying body is one ∧ everything else is produced by means of density and rarity.
- The underlying body is both one and many ∧ everything else is separated out from the indeterminate mixture.

Even if τὸ ὂν σῶμα τὸ ὑποκείμενον is not a natural phrase, there are some arguments in favour of the reading supported by the transmitted text. First of all, the text at lines a12–13[3] is not only transmitted by all Greek manuscripts and all ancient commentators, but also by the Arabic translation.[4] Furthermore, it is also confirmed by the best manuscripts

[3] a12–13: οἱ μὲν γὰρ ἓν ποιήσανες τὸ ὂν σῶμα τὸ ὑποκείμενον.

[4] This translation (ninth century AD) is an important witness to Aristotle's text and must be considered in any future critical edition. The history of the Arabic translations of the *Phys.* and of the ancient commentaries is reported in the bio-bibliographical catalogue of Ibn al-Nadīm (d. 995), the *Fihrist* (cf. Ibn al-Nadīm 1871–2 e 1970). Aristotle's *Physics* was translated several times during the ninth and tenth centuries. It was studied and commented on by many

(E and J2) and the Arabic translation at lines a21–2, which correspond to lines a12–23.[5] From a philological point of view, therefore, the reading supplied by the transmitted text should be preferred.

Moreover, this text seems to give a clearer understanding of Aristotle's general strategy in chapters 2–5. If we admit the transmitted text, we should translate the four lines as follows:

> Accordingly, making what is a single body, i.e. what underlies – either one of the three, or something else which is more solid than fire but less solid than air . . . (187a12–13)

> . . . the pairs of opposites are separated out from the one in which they already were, just like Anaximander says, and also those who say that the things that are are both one and many. (187a21–22)

According to this text, 'being (i.e. what is)' (τὸ ὄν) and 'beings (i.e. the things that are)' (τὰ ὄντα) are the subjects of the two sentences. Aristotle explains that his predecessors differ in that some of them (i.e. Ionian physicists, including Diogenes of Apollonia) think that being (i.e. what is) is one *and* that it is the underlying body from which everything else is derived by condensation and rarefaction; others (among whom were Empedocles and Anaxagoras) believe that being (i.e. what is) is one and many – that is the reason why it is more appropriate to speak of *beings* (i.e. the things that are) – *and* they believe that everything else that comes from those beings, considered as the principle, comes to be by separation. Following this reading, the 'principle' of the pluralists is not only the pre-existent *migma* from which beings come to be periodically (according to Empedocles) or only once (according to Anaxagoras), but generally speaking *that from which* beings come to be from other beings.

philosophers, including al-Fārābī (*c.* 872–950), Yaḥyā ibn 'Adī (d. 974), Abū 'Alī Ibn al-Samḥ (d. 1027), Ibn Sīnā (Avicenna, 980–1037), Ibn Bāǧǧa (1080–1138) and Ibn Rušd (Averroes, 1126–98). Greek commentaries by Alexander of Aphrodisias, Themistius and Philoponus were also translated into Arabic and used by all these thinkers (for a detailed presentation see Peters 1968). Ibn al-Nadīm mentions four partial or complete translations: the first one by Sallām al-Abraš; two incomplete translations by Qusṭā ibn Lūqā and al-Dimašqī, and the one carried out during the ninth century by Isḥāq Ibn Ḥunayn (d. 910). This is the only translation that has been transmitted to us by a unique manuscript (the Leiden manuscript: *Or.* 583), which reports in the margins some variants of the other translations, as well as the commentaries of some of the philosophers just mentioned. This translation was the one used by Averroes. It was edited by A. Badawī (cf. Badawī 1964–5) and translated into Latin together with Averroes' commentary by Michael Scotus (this is the second translation that precedes Averroes' commentary in the Junta edition (cf. Averroes 1962).

[5] a21–2: καὶ ὅσοι δ' ἓν καὶ πολλά φασιν εἶναι τὰ ὄντα.

According to this text, we should have the following division:

- Being is a single body, which is what underlies ∧ everything else is produced by means of density and rarity.
- Beings, considered as what underlies, are both one and many ∧ everything else is produced by means of separation from the indeterminate mixture (which must be considered as one and many at the same time).[6]

If we admit this reconstruction, the term 'being', as what pre-exists, must be primarily understood as a synonym of 'principle' and as designating 'what underlies' only in a derivative way. According to Aristotle's reconstruction, in all Presocratic doctrines, being as what is primary is what underlies, whether it be one or many, and it is, *as such*, principle of everything else, which derives from the principle.[7]

In support of this hypothesis, we may point out that the equivalence of being and principle is also in the background of chapter 2.[8] There, Aristotle explained that the inquiry into how many beings there are is ultimately an inquiry into the principle and the element, whether these are one or many.[9] This equivalence justified the relevance of the discussion of the Eleatic *aporiai* in the context of natural philosophy and more properly within the framework of *Physics* I. As Simplicius makes clear, even if the Eleatic monists were not speaking about nature, since they assumed that being is unchangeable, still Aristotle admitted that 'by treating being and the principle as identical', they were speaking about the principle.[10] For

[6] One can assume that in a sense Anaximander's doctrine occupies an intermediate position, since it supposes the existence of one single principle, i.e. the *apeiron*, while admitting that becoming is due to extraction. In another sense, however, one can point out that in so far as the *apeiron* contains everything within itself, the principle is no longer one, but many. This is how Philoponus understands Anaximander's doctrine in these lines: 'It is clear that in saying this Anaximander was undermining his own claims. For the common origin of all things, the intermediate, would no longer be one' (Philop. *In Phys.* 88.24–89.2).

[7] For an analogous reason, Averroes explains that the term 'being' is used in these lines as a synonym of principle. He claims that all Presocratic physicists admit one single 'being' that is also a substrate, even if according to some of them this underlying principle is actually and potentially one, while according to the others it is actually one and potentially many. He speaks of those thinkers as the partisans of the theory of *latentia* (Averroes, *LC Phys.* f. 21 E–H).

[8] See Clarke in this volume.

[9] *Phys.* I.2.184b22–5: 'And those who inquire into the number of the things that are are also inquiring in a similar way. For they inquire primarily into what the things that are are from, asking whether these things are one or many, and if they are many, whether there are a limited or unlimited number, so that they are inquiring into whether the principles and the elements are one or many.'

[10] Simpl. *In Phys.* 149.1.

this reason, it is necessary to examine their doctrines, even if their respective refutations do not pertain to natural philosophy.

Whatever reading one may choose, however, one must admit that a new element is introduced in the division of chapter 4 with respect to the one presented in chapter 2. While in the classification introduced in chapter 2, Aristotle pointed out two distinguishing features of the different schools, namely the number of the principles and their nature, i.e. changeable or unchangeable, homogeneous or heterogeneous, in the division of chapter 4 he distinguishes Presocratic philosophers according to the way in which they explain the coming to be of multiplicity out of what underlies.

In so doing, Aristotle emphasises the distinction between what underlies and the opposites and makes clear that in analysing Presocratic doctrines the question is not simply to find out how many principles of being they assumed, but to understand how their notion of principle can account for the coming into being of multiplicity: how Presocratic physicists 'produce everything else' (τἆλλα γεννῶσι: a15) from the substrate they considered as the primary being that is the principle of everything else. Despite the textual difficulty, this is a first important point to keep in mind: on either reading, Aristotle aims at refining and indeed modifying the division presented in chapter 2 by introducing a new criterion of subdivision, namely the different ways in which Presocratic theories of substrate can account for the coming to be of multiplicity.

With these considerations in mind, let us now take a closer look at the division. Aristotle claims that the thinkers who admit the existence of changeable principles differ in that some of them consider the coming to be of multiplicity as an effect of rarefaction and condensation of what underlies, while others consider it as the product of a separation. This consideration leads him to argue that the members of the two groups maintain that opposites are principles.

Concerning the first group, this conclusion necessarily ensues if we admit that condensation and rarefaction are principles of everything else and that they are opposites. At least, Thales, Anaximenes, Heraclitus and in particular Diogenes of Apollonia are included in this group, since each of them regarded one of the simple bodies (water, air or fire) as that from which everything comes to be by densification and rarefaction.[11] Aristotle mentions that some other thinkers also believed[12] that this unique primary

[11] At *Metaph.* A 8.989a6–9 Aristotle remarks that none of the early philosophers regarded earth as what underlies.

[12] For a detailed discussion of the different possibilities of attribution, see Ross 1936: 482.

body is an intermediate substance between fire and air ('something else which is more solid than fire but less solid than air').[13]

According to most ancient commentators,[14] Anaximander should also be included in this sub-group. But this is hardly plausible, since afterwards Aristotle explicitly classes him among those thinkers who consider the substrate (ὑποκείμενον) as being at the same time one and many.[15] Aristotle argues that Plato too should be considered as a member of this first group, since he believes that opposites (great and small) are principles, even if he considers what is one as the form and opposites as the matter, while the others consider 'the one underlying thing' as the matter.[16] The extent to which this is correct will be discussed by Lennox in this volume.

Concerning the members of the second group, things are less clear; it seems that Aristotle assumes that, according to them, opposites are principles in so far as they are already present in the original substrate. Empedocles must be included in this group, since he claims – at least according to Aristotle's interpretation – that everything emerges from the state called *Sphairos*, in which opposites are united by love and from which they emerge under the action of strife.[17] For the same reason, Anaxagoras must be included in it as well, since according to him homeomerous bodies, or more appropriately *spermata*,[18] and opposites[19] were indistinguishably mixed, before *Nous* separated them out and organised them.

On the basis of this consideration, Aristotle claims that according to all Presocratics, opposites are principles. This, however, does not imply that he wants to deny that according to them 'what underlies' is a kind of principle too. Since 'what underlies' is *that from which* everything else is derived either by condensation/rarefaction or by separation, it must be concluded that, according to those theories, what underlies is also a principle of change.

If we adopt this interpretation, the ultimate conclusion that Aristotle draws from his doxography is that, according to all Presocratics, opposites are principles, but they are not the *only* principles.[20] To be sure, Aristotle

[13] *GC* I.10.328b35, 332a21; *Metaph.* A 7.988a30. Elsewhere he talks about a substance intermediate between water and air (cf. *Phys.* III.4.203a18, 5.205a27; *DC* III.5.303b12; *GC* II.5.332a21; *Metaph.* A 8.989a14.), or fire and water (cf. *Phys.* I.6.189b3).

[14] Themist. *In Phys.* 13.18; Philop. *In Phys.* 87.1 (but cf. 88.24–89.2); Alex. *In Metaph.* 60.8.

[15] See n. 22.

[16] This remark confirms that the focus of Aristotle's discussion is less on being than on a unique something that underlies everything else. In other words, Aristotle is more interested in understanding the notion of 'substrate' than the notion of 'being'.

[17] B6. [18] B4, 10. [19] B4, 8, 10, 12, 15.

[20] On this particular point, see also Judson in this volume.

will explain later that the sense in which his predecessors understood the notion of ὑποκείμενον is an equivocal one, and that it is the 'first vague universal' – to use Aristotle's terminology in *APo* II.19 – from which our inquiry must begin. Properly speaking, as Aristotle will explain later, we cannot talk of this principle as a real 'substrate'. For *that from which* Presocratic physicists originate the plurality of things is inseparable from its affections and it is, for this reason, identical to the opposites.[21] Still, according to Aristotle's classification it is clear that 'what underlies' is also a sort of principle for the members of the two groups, and not just the opposites. I shall come back to this point later.

Aristotle classes at least two Presocratics in the second group: Empedocles and Anaxagoras. He goes on to explain that they differ from each other for two reasons: the former believes (a) that the process of separation happens cyclically and (b) that the other things that result from this process (the τἆλλα of line a23) are the so-called elements; while the latter claims that (¬a) the separation from the one/many-substrate happens once in the history of the universe, and (¬b) that what results from it is an infinite number of homeomerous bodies and opposites.[22]

Concerning Anaxagoras, in the next section the review of his arguments spells out that the infinite numbers of homeomerous bodies and opposites admitted by him are not just the results of the separation, but also, in more than one sense, the principles of this process, together with the *Nous*. As a matter of fact, not only were those infinite opposites and homeomerous bodies 'all together' before *Nous* separated them, but they are also actually present in everything that perishes and comes to be in our present world. In his reconstruction of Anaxagoras' arguments, notably the first one, Aristotle seems to give a minor role to this distinction. The second sense, on the other hand, seems to be relevant to the second argument.

II. Anaxagoras *apud* Aristotle

This is not the place for a full discussion of Anaxagoras' own doctrine. I will try only to go through Aristotle's account of Anaxagoras' arguments in order to understand which of his tenets Aristotle aims to rebut. This will allow us to understand the positive goal of his entire survey.

21 For more detail on this hypothesis see Cerami 2015: 63–88, in which I argue that the ultimate reason for the Presocratic mistake, according to Aristotle, is that they failed to perceive the existence of different meanings of the expression 'coming to be from'.

22 This passage confirms that Anaxagoras falls within the last group identified in ch. 2 (184b22).

Aristotle considers Anaxagoras' doctrine as grounded on two main arguments (at lines 187a26–31 and at lines 187a31–b1 respectively), from which he infers his general conclusion (187 b1–7). The first argument is based on an assumption that Anaxagoras shares with other physicists:

(1) 'no*thing* comes to be out of no*thing*'.[23]

From this statement, Anaxagoras assumes as an implicit corollary that:

(2) every*thing* must come out of every*thing*.[24]

He then infers his first conclusion that:

⊢ homeomerous bodies and opposites are infinite.[25]

From the same assumptions ((1) and (2)), Aristotle seems to attribute to Anaxagoras a supplementary statement:

(3) things were all together.[26]

This conclusion too stems directly from (1) and (2): if everything comes out of everything, things have always been together. The final lines of the argument (a29–30) add that for the same reason someone could identify 'coming to be a certain sort of thing' (τὸ γίγνεσθαι τοιόνδε) with alteration, while others regard the same process as an aggregation and dissolution. This last sentence raises some difficulties that I shall take into account shortly. Before turning to them, I would like to offer a general assessment of the argument.

Assuming the first statement, according to which 'what is' cannot come to be from 'what is not', Anaxagoras not only agrees with all Presocratic physicists that there is some sort of being that has always been, but he also places himself in a dialectical position with regard to Eleatic doctrines, in so far as they are also based on the same assumption.[27] This proximity with Parmenidean criteria of reality is not accidental.[28] Anaxagoras' commitment to a framework denying the possibility of an absolute generation is

[23] Which must be interpreted as 'a being cannot come to be from a non-being', or as in our translation 'no*thing* comes to be out of what is not'.

[24] That means 'every*thing* of the same form'.

[25] a26–9: 'Anaxagoras made <homeomerous bodies and opposites> unlimited in this way because he accepted *as true the general opinion of the physicists that* nothing *comes to be out of* what is *not*' (my emphasis).

[26] a29–30: 'this is the reason why they claim "things were all together"'.

[27] For a different reading of Anaxagoras' standpoint with regards to Eleatic doctrines, see Bodnár in this volume.

[28] Curd 2004; Furley 1989: 47–65, 1992: 74–80, 2002: 119–26.

clearly revealed, as we shall see, by his emphatic denial that coming-to-be and passing-away are genuinely real. But the same commitment is also at the core of his cosmogony. According to Anaxagoras, the cosmos evolved from an all-pervasive cosmic mixture, when Reason makes the ingredients emerge from this mixture through the force of rotation. To avoid Eleatic *aporiai*, Anaxagoras claims that even before the discriminative action of Reason the entire unlimited totality of ingredients were all together (see B17).

The argument goes as follows. Suppose we can extract from the mixture the totality of some type of ingredients by a process of progressive division. If such a complete separation were possible, that would imply that there was a least amount of those ingredients, such that, once this amount had been removed from our hypothetical mixture, none of them would remain. If subsequently something of the same sort did come to be out of that mixture, that would imply that something could come to be out of non-being, which is postulated to be impossible.[29] Therefore it must be concluded that the separation will never come to an end and that principles are infinitely separable, and hence infinite *tout court*.

Thus, from the appeal to his predecessors' *endoxon*,[30] Anaxagoras concludes first that 'things were all together' (B11–12) and then that homeomerous bodies and opposites were – as he claims – 'unlimited *in this way*' (ἄπειρα οὕτως). But what does this latter remark mean? And what kind of infinity is this argument supposed to prove? According to the proposed reconstruction, the argument reaches its conclusion (i.e. that something comes to be from nothing), even if the infinity considered is only quantitative and not qualitative. In fact, it is hard to see how this argument could prove that ingredients are infinite not only numerically, but also in form and species. Anaxagoras could infer the infinite number of kinds of homeomerous bodies and opposites if and only if he postulates an infinite actual number of those kinds in the present world. However, this argument does not seem to insist on that point.

[29] Suppose, for example, that there was a complete separation of the colour white from the mixture. Once the final minimal bit of whiteness had been removed from that mixture it no longer contains any whiteness at all. Then it would not be white. If subsequently something white did come to be out of the mixture that contains no whiteness, the principle that a being cannot come to be from a non-being would be violated.

[30] The identification of the assumption that 'no*thing* comes from no*thing* (i.e. what is not)' and the correlative one 'a being must come to be from a being' as *endoxa* recalls the passage of the *Sophist* in which the definition of nature, according to which it is 'that from which everything comes to be', is defined as 'popular' (265c).

In the next argument, as in his refutation, Aristotle will make clear that Anaxagoras is concerned to refute the possibility of arriving at a minimal size for the various kinds of *spermata* postulated by his theory, but not to prove that those kinds are infinite in number. In none of Aristotle's counter-arguments does qualitative infinity imply an infinite number of kinds of principles. Rather, the qualitative infinity of his principles is taken to require that they are all present in every mixture in a way that allows for bits of each to be extracted without limit. In this sense, as Simplicius suggests, it is more plausible to assume that, according to Anaxagoras, principles are undetermined in number, but determined in forms. Consider now the sentence at lines a30–1, and the whole clause put in brackets by Ross:

> This is the reason why they claim 'things were all together', and why he makes coming to be a certain sort of thing (τὸ γίγνεσθαι τοιόνδε) alteration, while they talk about aggregation and separation.

Ancient and contemporary commentators have tried to identify the thinkers to whom Aristotle alludes in this passage. Many if not most contemporary scholars suggest that Anaxagoras is the one who identifies 'coming to be a certain sort of thing' with alteration, while Empedocles and his partisans are those who unequivocally define generation and corruption as a consequence of aggregation (σύγκρισις) and separation (διάκρισις). The phrase τὸ γίγνεσθαι τοιόνδε designates what Aristotle would call absolute generation,[31] but it is hardly Anaxagoras' own terminology.[32] However, we may interpret it in the light of *Tim.* 49d–e. Here Plato explains that sensible things, which are in a permanent state of change, ought to be called 'of this kind' (*toiouton*), while what remains always the same must be said 'this' (*touto*). In this passage, Plato makes it clear that what we perceive as a product of a genuine generation is more properly a sort of modification of 'what receives all things', and this is what becomes of one kind or another according to the different forms that occur in it.[33] To be sure, Aristotle does not wish to equate Anaxagoras' account of sensible generation with the one put forward in the *Timaeus*.[34]

[31] This is the reason why I translate 'coming to be a certain sort of thing'. This expression should not be interpreted as a synonym of alteration. Rather, what Aristotle means is that something could change from a certain state to another so as to become another kind of thing.

[32] Simplicius, who has Anaxagoras' text in front of him, doesn't find this phrase in it (*In Phys.* 163.9–28).

[33] For a more detailed analysis of this text and other Aristotelian passages which support the claim that the expression τὸ γίγνεσθαι τοιόνδε designates substantial generation, see Cerami 2005.

[34] The two doctrines are in a sense contrary to each other, especially because in the *Tim.* what remains always the same is assumed to be as essentially deprived of any form. Anaxagoras' differences, as we shall see, are on the contrary inseparable from their matter.

Nonetheless, he would admit that for both philosophers the expression τὸ γίγνεσθαι τοιόνδε stands for a process that is loosely considered to be generation, although more properly it is a sort of alteration.[35]

That said, one could object to the hypothesis that it is Anaxagoras who equates generation with alteration by arguing that earlier, at a23, Aristotle had taken Anaxagoras to regard coming to be as the result of separation (ἐκκρίνουσι) from the *migma*, and not as the outcome of an alteration. Moreover, later in the chapter (188a5, a10), Aristotle explicitly affirms that this separation is the result of separation (διάκρισις). The authenticity of this doctrine is not questionable. It is confirmed by several fragments in which Anaxagoras claims that the processes loosely called generation and corruption are actually association and separation (notably B17). Adducing the same argument, Simplicius informs us that Porphyry suggested the identification of the unnamed supporter of the assimilation of generation to alteration with Anaximander.[36]

The hypothesis that this passage does not allude to Anaxagoras could be supported further by the suggestion of Heidel as reported by Diels (1912, p. 388, nn. 25–6). Heidel regarded the whole clause in parentheses (a29–31) as an explanation meant to be inserted after τἆλλα at a23. According to him, however, the author in question is Anaxagoras. But it is difficult in this case to identify the others who 'talk about aggregation and separation'.[37]

Contrary to this suggestion, one can retort that there is another passage that lends supplementary weight to the evidence in favour of the first hypothesis. The idea that Anaxagoras makes generation alteration seems to be confirmed by what Aristotle says about Anaxagoras in *GC* I.1. In this chapter, Aristotle criticises Anaxagoras, as well as Empedocles and Democritus, for having failed in their attempts to offer an adequate and coherent explanation on the basis of what they have defended, namely the existence of alteration and therefore of generation. He puts forward a single argument against all those thinkers who assume the existence of a 'plurality of matter' ('those who make matter more than one': 314a11) *and* who try to account for the existence of alteration. He explains that, due to the fact that they posit a plurality of principles, alteration would be an inexplicable phenomenon for all of them[38] (I will come back to this general

[35] For the same reason Aristotle puts Plato among those thinkers who admit the existence of one principle and two opposites. The classification of *GC* I.1 according to which the same thinkers must consider generation as alteration is in the background.

[36] Simpl. *In Phys.* 163.16–18.

[37] The οἱ δὲ σύγκρισιν καὶ διάκρισιν at a31.

[38] *GC* I.1.314b15–25.

argument later). At 314a6ff., Aristotle adds that Anaxagoras deserves a harsher judgement, since 'ignoring the proper meaning of his words', he assimilates generation to alteration. For this reason, he has to be blamed not only for having been incapable of explaining phenomena, but also for having failed to draw the necessary conclusions from his own premises.

In light of this passage of *GC* I.1, it seems quite plausible that the thinker Aristotle alludes to in 187a30 is Anaxagoras. We can naturally assume that, according to Aristotle, there is no inconsistency in regarding Anaxagoras' account of generation to be an account of both alteration and separation. By interpreting his account of generation as an alteration, Aristotle would take the ('mistaken') point of view of Anaxagoras himself, so as to distinguish him from Empedocles; by insisting on the identification of generation with separation, Aristotle would make clear that the kind of process admitted by Anaxagoras cannot be considered as a genuine alteration. From this point of view, Empedocles and Democritus should also be included in the aforementioned refutation.

*

The second argument that Aristotle attributes to Anaxagoras is based on the same series of premises (1)–(3) together with the supplementary assumption that 'opposites come to be out of one another', which is assumed to be something based on what is seen.[39] The aim of the argument is to demonstrate that opposites must be together. The argument has the same structure as the first one. If (1) 'no*thing* comes to be out of no*thing* (i.e. what is not)', we must assume that (2) every*thing* must come out of every*thing*, and therefore that (3) things were all together. Since 'opposites come to be out of one another' – which is another *endoxon* that Anaxagoras accepts as true – the two opposites must pre-exist and must be present together. Even if we could distinguish them, one opposite must always imply and include the other. If this were not the case, something could come to be out of what is not. Thus, it must be assumed – concludes the argument – that 'everything comes to be out of things that are and that are present as constituents, but cannot be perceived by us in view of the smallness of their bulk'.

Ingredients, both homeomerous bodies and opposites, are infinite in so far as their separation can be pursued indefinitely; but precisely in view of

[39] 187a31–b7.

the infinite smallness of their size, they cannot be perceived *by us*. This final clarification confirms that the same is true both of the original *migma* and of our present world: opposites are all together. Anaxagoras is asserting that even when a stuff or quality is not apparent, it must still be present, just as in the original state.

This piece of doctrine is confirmed by several of Anaxagoras' fragments. In fr. B1, the ingredients in the original mixture are characterised in exactly the same way:

> all things were together, unlimited in extent (*plethos*) and in smallness, for the small, too, was unlimited.

Elsewhere (B17) Anaxagoras makes clear that the same is true of our present world as of the pre-cosmic state:

> The Greeks are wrong in talking about coming-to-be and perishing; for nothing comes to be or perishes. But <everything> is mixed together and separated from the things that are. And thus they would rightly call coming-to-be being mixed and perishing being separated.

Anaxagoras posited the opposites and the basic entities called 'seeds' as his principles (ἀρχαί) not only in order to claim that they are what existed temporally prior to everything else, before the ordered world arose, but also to argue that the actual world must be explained through these ἀρχαί. The actual world could not have arisen without them. In this sense, they are the genuine ἀρχαί that account for the phenomena of the sensible world.

The denial of 'a smallest' portion is clearly connected to the assertion that what is cannot not be. The process loosely called generation is not a coming out of what is not, but a change in the proportions of what is. Ingredients are always produced by concretion of smaller portions of themselves which were present in substances predominantly of another kind (B10).

The reality of change is not denied, but it is exclusively granted to this kind of transformation, namely the transformation by separation of a principle that is at the same time one and many. From this point of view too, Anaxagoras replies to Parmenides' *aporiai* assuming that each thing is in a sense genuinely one. The impossibility of stopping the process of separation entails that everything is constituted always and at the same time by *all* the same infinite principles.

To conclude the analysis of Anaxagoras' arguments, two points must be stressed. First, in neither of these two arguments does Aristotle tell us explicitly which kind of infinity is taken into account. In the second argument, he seems to consider that only the quantitative nature of the principles is involved.[40] Second, Aristotle makes clear that according to Anaxagoras the ingredients are not just the opposites and the seeds, but also the homeomerous bodies and the opposites.[41] All of them are equally involved in the demonstration.

The conclusion of the section confirms our views as to which of Anaxagoras' tenets Aristotle is drawing attention to. Note that the claim that *Nous* is the principle of order is not mentioned. Aristotle restates the general conclusion of the two arguments, namely that everything is mixed in everything, because everything comes to be from everything; and concludes that, according to Anaxagoras' supporters, things only look different. Even if things are said to be 'one thing rather than another', they are the same, since they include all the same infinite principles. They differ only in view of what is numerically predominant in the mixture.

The final lines lead to the central topic. The tenet at stake for Aristotle is that no actual separation of those infinite homeomerous bodies and opposites is possible; no whole could be purely such or such a thing. If we suppose that an object can be actually affected by just one quality, or that it can lose it, then that quality would be separable and possibly non-existent. Opposites are neither pure qualifications nor predicates; as the seeds, they are intrinsically qualified matter. According to Aristotle's reconstruction, Anaxagoras' principles, homeomerous bodies and opposites alike, are types of matter, not a separable matter modified by different qualifications. Substrate, in other words, is necessarily tied to all infinite qualifications whose different proportions generate plurality. This will be the crucial tenet at stake in Aristotle's refutation.

III. Aristotle's Refutation of Anaxagoras' Doctrine

All the arguments contained in lines 187b7–188a5 are intended to prove that the principles present within a single finite being cannot be infinite from a quantitative and/or a qualitative point of view. In none of these arguments does qualitative infinity imply an infinite number of kinds of principles. As noted above, what is at issue is only the co-existence of all of them in all mixtures in such a way that their extraction from any given mixture never results in any mixture lacking any of them.

40 Simplicius, as we have already seen, argues that Anaxagoras took for granted only the numerically infinite, while he claimed that principles are finite in forms and kinds.

41 In favour of the first hypothesis, see Schofield 1980; Tannery 1930; Vlastos 1950. For a defence of the idea that ingredients include opposites and seeds, see Graham 1994.

An outline of each argument is in order before pointing out their common assumptions. First, I shall examine the five arguments that Aristotle puts forward in lines 187b7–188a5; then I shall consider the final passage (188a5–18) in which Aristotle draws some positive conclusions that reveal two supplementary difficulties implied by Anaxagoras' theory.

First argument: 'if the whole is knowable, the parts are knowable'.[42] The first lines of the argument are meant to characterise that which is *apeiron* as such: the unlimited as such is unknowable. If Anaxagoras' principles are unlimited from a quantitative as well as from a qualitative point of view, namely according to number and size as well as according to form, then that which is constituted by those principles is not knowable. The conclusion follows if and only if we assume (i) that things must be knowable; (ii) that the whole is constituted by its parts; and (iii) that there is no other respect in which an *apeiron* could be known other than quantitatively and qualitatively.

Second argument: 'if the whole is limited, the parts are limited'.[43] The argument goes as follows: if the parts were unlimited according to size, the whole would be similarly unlimited. But no living being in nature, animal or plant, can be indefinite according to size, so neither can any of its parts. Aristotle makes it clear that according to this argument we must assume that the whole under consideration is a living being, and that the constitutive principles are its homeomerous parts, such as flesh, bones etc.

Third argument: 'everything cannot be in everything'.[44] The argument is partially based on the conclusion of the preceding one: there are parts, like flesh or bones, which have a certain limit to their size. It is used here to refute one of the assumptions implied by the existence of infinite principles, namely that everything is present in everything. The argument goes as follows. Consider two uniform bodies in the *migma*, e.g. water and flesh; assume also that flesh can be exhausted by a sufficient number of subtractions of finite parts from it. If, as the second argument has already proved, flesh has a definite minimal size, then we must assume, at a certain point, that we cannot separate a part smaller than the minimal one. In this case, only two options remain: either the separation comes to an end, in which case water remains without flesh, with the consequence that not everything is in everything; or else the separation goes on forever, in which case there will be an unlimited number of parts of flesh of the same limited size in a limited magnitude; but that is impossible.

[42] *Phys.* I.4.187b7–13. [43] *Phys.* I.4.187b13–21. [44] *Phys.* I.4.187b21–34.

Fourth argument: 'everything does not come to be out of everything'.[45] I suppose that this argument is ultimately meant to refute the other tenet stated in the preceding argument (187b24): 'anything can come to be out of anything'. This argument too is based on the second argument's conclusion: flesh has minimal size. If every body must become smaller by the subtraction of some smaller body of the same kind *and* if flesh – as we have already seen – has a minimal size, then from the minimal size of flesh nothing can come out; for there would be a smaller quantity of flesh than the smallest one.

Fifth argument: 'unlimited bodies imply an infinite number of entities actually separated, i.e. existing, in each uniform part, and then again in each of these infinite entities infinite others'.[46] Consider – contrary to the second argument's premise – that bodies are unlimited, so as to contain an unlimited amount of all the 'seeds'. In this case, we must postulate an infinite number of real parts that are separated from one another, each one of which contains an infinite number of entities, and then again in each one of these infinite entities there are infinite others, and so on *ad infinitum*. But this is nonsense, since it means that the infinite is multiplied infinitely many times.

*

Consider now the common tenets of Aristotle's refutation. First of all, we can gather that the ultimate aim of all these arguments is to prove that those parts of the composite whole that are *archai* cannot be infinite either according to their number, or according to their size and species.

All arguments are grounded on two fundamental assumptions: (I) the whole consists of parts; (II) what is true of the whole composite must be true of the parts that are principles. Since the parts and the whole are essentially dependent upon each other, we can infer conclusions concerning the parts that are principles from the assumptions concerning the whole: if the whole is knowable, the parts that are principles are knowable; if the composite whole is definite in size, the parts that are principles are definite in size. All arguments assume that, in this respect, what is true of the whole composite must be true of the parts that are principles. This assumption constitutes the core of Aristotle's refutation and confirms at the same time a number of fragments in which Anaxagoras acknowledges that this 'analytical'

[45] *Phys.* I.4.187b35–188a2. [46] *Phys.* I.4.188a2–5.

method, which goes from 'what is known' to 'what is unknown', is the path that leads human reason to a certain knowledge of the sensible world.[47]

From this assumption Aristotle infers the fundamental principle that he opposes to Anaxagoras' doctrine: constitutive parts that are principles must be definite in number and size. In natural things, separation does not go *ad infinitum*: the separated body at any time will still not exceed some magnitude in smallness.

As Charlton (1970, ad loc.) makes clear, Aristotle's claim can be interpreted in two ways: (i) for all x, x is a separated part implies that there is a y such that x is not smaller than y; (ii) there is a y such that for all x, x is a separated part implies that x is not smaller than y.

The second interpretation of Aristotle's claim is the basic assumption of the theory of *minima naturalia*: all natural beings have a definite limited size. Flesh, bones, blood and all homeomerous parts have by nature an amount of matter beyond and below which the form cannot instantiate. Things are made up or composed of *minima*, which are in some theoretical sense the smallest thing possible for each natural body, even if they are not necessarily well-distinguished particles.

This is not the place to consider this 'neo-Aristotelian' piece of doctrine.[48] Suffice it to remark that we find hints of this theory in ancient commentators (Alexander, Themistius, Simplicius and Philoponus), while we read a more radical version in most medieval Arabic and Latin philosophers who follow the Peripatetic tradition.[49] I shall mention two examples, since they are extracted from ancient Greek commentaries on *Phys.* I.4.

In his paraphrase of *Phys.* I.4 Themistius claims that we must assume a minimal size of every natural body, in order to avoid all the *aporiai* arising from the assumption of their infinite separability. According to one of them, Themistius concludes that it will no longer make any difference to speak of more or fewer constitutive parts of different beings when the size of those basic principles is indeterminate both as to greater and as to lesser amount.[50]

[47] Two fragments in particular attest that Anaxagoras considered the whole thing as knowable from a human point of view (B21) and as actually known by the *Nous* (B12). Concerning the use of analogy and visible signs to elucidate the hidden mechanisms of nature, see B21a.

[48] On this doctrine, see Cerami 2012 and the bibliography it contains.

[49] Those authors, more explicitly than Greek commentators, asserted that the theory of a *minimum naturale* concerns not only homeomerous bodies, but also simple bodies, and all affections.

[50] Themist. *In Phys.* I.4.15.25–31. I owe this quotation to Alessia Stesiano who is currently working on her Ph.D. thesis devoted to Avicenna's *Physics* and to his Greek and Arabic sources.

Commenting on the same passage, Philoponus makes the same claim.[51] Some pages earlier, in his commentary on the first argument (187b7ff.), he asserted that the forms naturally subsist in some finite quantity, and do not naturally grow to just any size, nor naturally shrink to just any smallness: there is a limit both to the greater and to the smaller, beyond which the form cannot exist.[52] Philoponus goes on to make clear that this is true of the forms of homeomerous bodies as well as of that of the elements: the form of none of them can occupy just anything whatever, either as regards largeness or smallness, 'but there is a limit, with regard both to the greater and less, beyond which the form cannot exist'.[53]

Even if the theory of *minima naturalia* was not unequivocally defended by Aristotle, there are some texts suggesting that a physical body cannot be infinitely divisible and still maintain its physical properties.[54] Aristotle seems to maintain the necessity of a *minimum naturale* in different contexts; in most of them this notion is used to explain the necessity, for every kind of change, of the existence of two boundaries.[55] In *Physics* III,[56] discussing the subject of the infinite by addition and by division, Aristotle suggests that the matter–form structure renders only the second kind of infinite possible, because matter is infinitely divisible, while the form is associated with the whole. In the context of the theory of substantial generation, Aristotle makes it clear that there is a limited size below which elements pass away and the ingredients of a mixture are mixed more readily.[57] All these passages should be considered as parallel to *Phys.* I.4 in so far as they all assume the necessity of stopping the separation of a natural body's components, and conclude that the principles extracted must be limited in number and in kinds.

It is important to note, however, that Anaxagoras' view concerning the quantity and size of principles is not the only aspect of his theory that is at stake in *Phys.* I.4. As I have already suggested, according to Aristotle it is also the intrinsic nature of Anaxagoras' principles that involves *aporiai* and must be refuted. The infinite number of Anaxagoras' principles necessarily goes together with their non-analysable nature. For Aristotle, Anaxagoras'

[51] Philop. *In Phys.* 103.33ff.
[52] Cf. Philop. *In Phys.* 481.3–6: 'The division of the form often comes to an end because it has a certain limit, with regard to the smaller as to the greater (for it has been said that forms are not naturally able to remain in every magnitude), but because of matter the cutting does not come to an end' (trans. in Edwards 1994: 143).
[53] Philop. *In Phys.* 96.26ff.
[54] For an almost complete survey of these texts, see Van Melsen 1960: 41–4.
[55] *Phys.* VII.5.250a20. [56] *Phys.* III.7.207a33–6. [57] *GC* I.10.328a23–8.

principles necessarily imply the impossibility of the separation between a real substrate and the properties that inhere in it. This is the fundamental mistake by Anaxagoras, according to Aristotle. It is not the plurality of principles in general that is in question, but the plurality of constitutive principles. To use Aristotle's terminology, Anaxagoras' mistake is that, like Empedocles, he makes 'matter more than one' (*GC* I.1.314a11), that is, a substrate that is both one and many (187a21–2), because affections are *in no way* separable from it.

The passage confirms that this is the doctrine at stake, namely the overall non-separability of Anaxagoras' principles. It is true – argues Aristotle – that separation does not come to an end, but this is not true in just any way.[58]

Aristotle grants that affections are not separable, but argues that Anaxagoras does not understand in which sense and why this is true. For, although he too holds that affections are not separable, he does not understand that this is true only if one means that they are not separable in an unqualified way, which is being inseparable from a substratum. If affections were separable – Aristotle argues – we would have 'a pale or a healthy which is nothing else'. As Aristotle himself makes clear in *Metaph.* Z 1.1028a20ff., 'walking', 'being healthy', 'being seated' and all affections neither exist per se nor can they be separated from substance. 'What walks', 'what is healthy' etc. exist only because there is a substrate receiving the affection. From this point of view, Anaxagoras is right, since he states that complete separation never will take place. But precisely for that reason, we have to distinguish between substrate and affections. For if this separation were not possible, there will be 'a pale or a healthy which is nothing else, not even <an attribute> *of* an underlying thing'. If we suppose, as Anaxagoras does, that separation is not at all possible, we might conclude that principles are all intrinsically one, so that each constitutes an absolute unity incapable of any change and, contrary to what Anaxagoras himself would conclude, completely separable.

The same mistaken assumption is also at the basis of Aristotle's criticism in *GC* I.1. As we have already seen, in this chapter Aristotle blames all 'pluralists', because while they distinguished generation from alteration, they were actually incapable of accounting for alteration. If they drew the right conclusions from their own premises, they might conclude that alteration is impossible. But none of them denied the existence of alteration.

Anaxagoras, Empedocles, and even Democritus, to some extent, are accused of having committed the same mistake: admitting the existence of

[58] *Phys.* I.4.188a5–18.

alteration, and claiming at the same time that there are many principles. The first of these, Anaxagoras, is held to be 'guiltier' than the others for having claimed, despite his premises, not only that alteration exists, but also that generation *is* alteration.

Some contemporary scholars dispute the claim that Aristotle's assumption (according to which pluralists, following their own principles, might distinguish alteration from generation and deny its existence) is based on some theoretical arguments.[59] They suggest that this claim is nothing but a historical remark. Nothing *a priori* compels them to distinguish the two phenomena,[60] and this distinction is just an assumption of a certain group of them, namely Anaxagoras and Empedocles.

One may reply that, even if it is true that Aristotle attacks a certain kind of historically attested pluralism, his conclusions are grounded in some theoretical assumptions. Aristotle attributes to Empedocles a theory according to which alteration is impossible, because the material bodies that constitute our sensible world are essentially defined by their proper specific *differentiae*, and because they are incapable of transforming into one another.[61] This reconstruction is based on two assumptions attributed to Empedocles' partisans: (i) elements are constituted by their affections, which function as specific differences; (ii) elements are incapable of transforming into one another. Empedoclean elements, according to this reconstruction, cannot lose those affections without completely vanishing. Deprived of their function as underlying body, they are completely identified, from Aristotle's point of view, by their primary qualities.

If we assume, as Aristotle does,[62] that alteration is ultimately produced by a transformation of elementary primary qualities, then a theory implying the non-transmutability of those qualities can never account for the existence of a qualitative change. Thus, Empedocles' mistake is to make his elements intrinsically determined by a number of irreplaceable affections. This assumption constitutes the grounds for Aristotle's claim that Empedocles makes 'matter more than one'.

Although Anaxagoras is not Aristotle's primary target in *GC* I.1, nonetheless the above argument engages his doctrine too, since Anaxagoras, just as much as Empedocles, admits a plurality of material principles. For the same reason given for Empedocles' failure, Anaxagoras misunderstands

[59] Cf. Williams 1982: 61; Brunschwig 2004.

[60] Williams 1982: 61ff. remarks that there is no theoretical argument that compels pluralists *in general* to distinguish alteration from absolute generation. Only the pluralists considered by Aristotle are obliged to assume this distinction on the basis of their doctrine.

[61] *GC* I.1.314b15–25. [62] *GC* II.4.331a7–11.

generation and is compelled to deny alteration because he finally has to conclude that we cannot separate affections at all from what underlies. Hence what underlies is incapable of receiving opposite affections, and opposite affections are incapable of properly being said *of* it. This assumption necessarily implies, for all pluralists, the impossibility of accounting for alteration and generation at the same time. Aristotle concludes that it is for this reason that we must assume the existence of one substrate for the two opposites in all changes, namely changes according to quality, quantity, place, as well as substantial generation.[63]

In *Phys.* I.4 Aristotle does not focus on the inability of Anaxagoras' doctrine to explain the natural phenomena linked to generation. Aristotle's aim is not to prove that Anaxagoras was unable to explain the existence of alteration and generation, but his goal is rather the more general one of showing that he cannot account for any change at all. Nevertheless, the basic assumption upon which his refutation is based must be the same: the inability to understand. In particular, Aristotle holds that both Anaxagoras and Empedocles were incapable of grasping the need to acknowledge a kind of separation between what underlies and its affections. Since Anaxagoras' underlying principles are intrinsically both one and many (that is, actually both one and many), he cannot account for any transformations in a sensible world.

Thus, in a very Socratic way, Aristotle concludes that Anaxagoras' *Nous* is an unreasonable principle, since it wants to produce a separation which it considers as impossible to produce! With respect to any affection, separation cannot be produced, since the substrate is ultimately identified by all the same infinite affections. With respect to quantity, the separation cannot be produced either, since the size of these underlying principles is undetermined.

As Aristotle makes clear in the final lines of *Phys.* I.4 as well as in *GC*, Anaxagoras also misunderstands the generation of homeomerous and simple bodies. Concerning homeomerous bodies, we can affirm that in one way they are divisible into what are always and in each case the same principles, but in another way we must admit that they do not so divide. For, at a certain moment they are divided into more basic bodies, e.g. in the way in which clay can be divided into, not clay, but water and earth. Moreover, concerning the same kind of bodies, we cannot claim that they are constituted and come to be out of one another, as we claim that bricks come from a house and a house from bricks. Bricks are actually present in a

[63] *GC* I.1.314b26–8.

house, and this is the reason why a house comes from bricks, namely that it comes from them as from its constitutive actual parts. Simple bodies, on the contrary, are not constituted by one another as by actual constitutive parts. To explain this entire process, we must admit a substrate that is separable in thought, even if it is always accompanied by opposites.

Thus, we must conclude that the acknowledgement of a plurality of underlying principles cannot but imply that it is impossible to separate the substrate from its affections. This is the same incorrect conclusion to which Anaxagoras, Empedocles, and also Democritus, are inevitably carried, and against which Aristotle will elaborate his own notion of substrate. From this point of view, the crucial mistake of all these thinkers is to think of opposites as identical with the underlying principle, given that they cannot separate affections from what underlies them.

As all the counter-arguments of *Phys.* I.4 show, the assumption of an infinite number of underlying principles is essentially connected with that of their inseparability. But as these arguments also show, Anaxagoras' doctrine is even more unreasonable than Empedocles', since at least Empedocles' doctrine assumes a finite number of principles. From this point of view, even if both of them miss the correct meaning of the notions of separation and of substrate, Empedocles' doctrine is in the end better than Anaxagoras'.

IV. Conclusions

To conclude, we should acknowledge that in *Phys.* I.4 Aristotle has two parallel aims. (i) First, by establishing his new doxographical criterion according to which Presocratic physicists differ in that they give two different accounts of the coming to be of multiplicity (i.e. alteration vs. separation), Aristotle intends to make it clear that all these thinkers hold that opposites are principles, though not the only principles. (ii) Secondly, by attacking the pluralists' doctrine, and especially Anaxagoras' theory of an infinite number of principles, Aristotle wants to show that separation is a condition needed for the explanation of change, but that this notion must be qualified. If we admit an infinite number of principles, it turns out that, contrary to what Anaxagoras himself believes, we cannot account for the existence of a plurality of knowable things. Separation cannot go *ad infinitum*. On the contrary, with a proper understanding of the notion, we can see that it must stop, due to the fundamental distinction between affections and the substrate to which they are attributed. This notion of separation, and not the Anaxagorean one, will be the core of Aristotle's own doctrine, as well as the basis of the assumption, to be established in the following chapters, that the principles are three in number: two opposites and one substrate. If in chapter 4 Aristotle shows that Presocratic physicists also assume a certain conception of substrate, it is in the following chapters that Aristotle will make clear in what sense we must rectify those notions in order that we may arrive at the 'scientific' notion of opposites and substrate. In *Phys.* I.5–6 Aristotle explains that the Presocratic principles must be analysed in order to distinguish in them 'what really underlies' from 'what – strictly speaking – changes'. In *Phys.* I.7, he will further make it clear that this division necessarily implies two meanings of 'coming to be from something' as well as two related understandings of the phrase 'what comes to be', of which one applies to what does persist, and the other to what does not. It is in this sense that Presocratic principles constitute the ὅλον, or the 'whole', and the '[confounded] things' that mark the starting point of our inquiry into principles. Including many things as their parts, these are the notions that must be analysed. *Physics* I.4 is a crucial part of the student's learning to come to see in them two 'parties', 'what is replaced' and 'what underlies'.[64]

[64] *Phys.* I.7.190a9–10: 'Of the things that come to be, in cases we describe as involving simples, in one case something comes to be by remaining, in the other by not remaining'.

Physics I.5

Lindsay Judson

Physics I.5 has three main parts: 188a19–30 picks up on part of what was said in chapter 4 – that Aristotle's predecessors in one way or another made opposites principles; 188a30–b26 considers the idea that opposites are principles ἐπὶ τοῦ λόγου; 188b26–189a10 returns to the subject of Aristotle's predecessors. It is important to bear in mind that chapter 4 did not ascribe to these predecessors the view that opposites are the *only* principles (see 187a12–26). As we shall see, this will be especially relevant to the argument at 188a25–30, to the whole of the argument ἐπὶ τοῦ λόγου, and to the three statements of the conclusion which Aristotle appears to have reached: 188b25–6, 188b28–30, and the concluding sentence of chapter 5 (189a9–10), 'that the principles must be opposites is evident'. Most of the key issues in I.5 concern either the chapter as a whole or the relations of some of its parts, and so I shall organise my chapter around these rather than by proceeding line by line. The issues are: (I) how we should understand, and best translate, ἐναντίον; (II) how we should understand Aristotle's various claims about principles (ἀρχαί); (III) the tension between the claims that ἐναντία are not from each other and yet come to be from each other; (IV) questions about the logical grammar and/or the ontological commitments of these claims, about what the principles in question are principles of, and about Aristotle's insistence that these principles are *opposites*; (V) the assessment of Aristotle's position in chapter 5. I conclude with (VI) a commentary on the final part of the chapter, 188b26–189a10.

I am grateful to the participants in the workshop for very helpful discussions, and especially to Laura Castelli, David Charles, Tim Clarke, Alan Code and Diana Quarantotto.

I. How Should We Understand and Translate ἐναντίον?

I shall discuss two questions: is Aristotle using ἐναντίον in a technical sense, and how should we best translate the term? Even if we suppose the answer to the first question is yes (I cast some doubt on this below), I prefer the translation 'opposite' to 'contrary'. The modern notion of being contraries is a purely logical or formal one, defined in terms of (in)compatibility relations;[1] this notion stems from Aristotle's treatment of contrary statements in the *De Interpretatione*.[2] Aristotle gives set-piece accounts of his technical notion of ἐναντία in *Cat.* 10 and in *Metaph.* I 4–10: it is perhaps unclear whether these two accounts yield the same sense or different ones.[3] But in either case Aristotle's notion of ἐναντία is not primarily about statements or their truth makers. It is clear from his discussion in *Cat.* 10 that as applied to properties or to terms it is not a purely formal one.[4] Thus he distinguishes ἐναντία from ἕξεις and privations on the grounds (i) that a privation cannot precede a ἕξις and (ii) that the acquisition of a privation is irreversible (so that privations acquired after the ἕξις has been possessed cannot be lost), whereas change into an ἐναντίον is reversible; (iii) that in the course of this he classifies essential properties which are the same as 'contingent' ἐναντία (such as fire's being hot) as themselves ἐναντία rather than as ἕξεις and privations, despite their inability to be lost (12b26–13a36). These are causal grounds, not grounds relating to formal (in)compatibility. Similarly, in *Metaph.* I 4 Aristotle distinguishes ἐναντία from contradictories on the grounds that the former but not the latter 'admit of intermediates' (1055a38–b3): given that not all ἐναντία do admit of intermediates,[5] Aristotle is clearly thinking of them as a *kind* defined independently of their (in)compatibility relations – and that is exactly how he proceeds in *Metaphysics* I, where he defines ἐναντία in terms of maximum or complete difference within a genus.[6] Finally, there is the idea advanced in *Metaph.* I 4 that intermediates are *composed of*

[1] So, e.g., 'Wffs (statements) *A* and *B* are contrary if they can't both be true but may both be false' (Haack 1978: 244; cf. Lemmon 1965: 69).

[2] *Int.* 10–14; cf., e.g., Lemmon 1965: 170; Mates 1965: 198.

[3] I am grateful to Diana Quarantotto for pressing this point. In what follows I shall remain neutral as to whether the *Categories* sense of ἐναντία is the same as the one developed in *Metaph.* I (for one issue see n. 12).

[4] Bogen 1991 reaches a similar conclusion by a somewhat different route.

[5] *Cat.* 10.11b38; *Metaph.* I 7.1057a18–19; cf. 188a36–b8, discussed below.

[6] Something like this point is recognised in some nineteenth-century logic textbooks in the distinction between contrary terms and contrary propositions: see Keynes 1894: 50 and 87. Note that elsewhere, in *Metaph.* I 7.1057b19–22, and in Δ 10.1018a25–31, Aristotle somewhat bafflingly says that some ἐναντία are not in the same genus: see Castelli 2018, ad loc.

ἐναντία.[7] For these reasons I prefer to avoid the translation 'contrary' even for Aristotle's technical usage(s), and use 'opposite' instead.[8]

Is Aristotle using ἐναντίον in a technical sense? In truth, I think that this is indeterminate. On the one hand, Aristotle is using the term to characterise the position of the Presocratics, who seem not to have used the term ἐναντίον in this sort of context at all,[9] and in any case certainly lacked Aristotle's technical usage. The opposites taken to be fundamental by one early physicist or another are, as Aristotle says, pairs such as hot/cold, wet/dry, dense/rare, full/empty; typically – though not in the case of full/empty – these are thought of as mutually excluding extremes on some scale or spectrum, and as in some way involving opposed causal powers:[10] this idea is not unrelated to Aristotle's technical usage, but it is hardly identical to it. On the other hand, it is of course common Aristotelian practice to characterise his predecessors' views in his own terms and thus to align their approaches and conclusions more closely with his own (or to oppose them more sharply) than might be historically warranted. There is some reason to doubt that Aristotle's usage of ἐναντίον is his technical one, however. As is well known, he is going to go on to claim that ἐναντία are principles because form and formlessness – or, more generally, form and privation – are principles. It is a familiar problem that form and privation are ill-suited to be ἐναντία,[11] and indeed that in *Cat.* 10, as we have seen, Aristotle distinguishes ἕξις and privation as a different species of ἀντικείμενον from ἐναντία.[12] What is most striking in our chapter is perhaps not so much Aristotle's doctrine as his terminology. After expounding the claim that (sc. in the case of incomposites (ἁπλᾶ)) all

[7] See p. 147 and n. 58.

[8] 'Opposites' is usually used to translate the more generic ἀντικείμενα (see *Cat.* 10; *Top.* II.8 and V.6; *Metaph.* Δ 10, Α 4 and 7): this is itself not a very helpful translation in any case, since one class of ἀντικείμενα, correlatives, such as knowledge and the knowable, are not typically opposites of any sort. For ἀντικείμενον the workshop group debated between 'set against each other' (which is too ungainly to be ideal) and 'opposed' (which has too close a connection with 'opposites' to be ideal), and decided to use the latter. In translating ἐναντίον, etc., in Simplicius and Philoponus I shall use *enantion*, etc., so as to avoid prejudging how they understand the terms.

[9] With the exception of Parmenides, who uses τἀντία at B8.55. I am grateful to Tim Clarke for reminding me of this.

[10] Cf. Bostock 1982: 189, and below p. 147; for a general discussion see Ferguson 1969 and Lloyd 1966.

[11] See, e.g., Simpl. *In Phys.* 182.2–5: '[Aristotle] himself, as we shall learn, in saying that the *enantia* are principles, will posit form and privation as the principal *enantiōsis*: these are set against each other not as *enantia* but in some other way.' See also Bostock 1982: 190–1. For Philoponus' partial defence of the view that they are *enantia* see below, pp. 150–51.

[12] The position in *Metaph.* Ι 4 is more complicated. On the one hand, it apparently accepts the same four species of ἀντικείμενον (ἀντίφασις, στέρησις, ἐναντιότης, τὰ πρός τι: 1055a38–b1); on the other, it characterises the 'primary ἐναντίωσις' as ἕξις and (complete) privation (see below p. 147).

changes are between ἐναντία, he says at 188b8ff. that the same is true for composites (σύνθετα):

> And this holds in the same way in the other cases, since the same account holds also for those of the things that are which are not simple but composite; but this is overlooked because the opposed conditions (τὰς ἀντικειμένας διαθέσεις) have not been named. For it is necessary for everything structured to come to be from unstructured and the unstructured from structured, and for the structured to cease to be into unstructuredness – and this is not just any unstructuredness but the opposed one (τὴν ἀντικειμένην).

The switch to ἀντικειμένας might seem to betray some recognition of the point that structured and unstructured are not ἐναντία – yet Aristotle draws his conclusion about ἀρχαί once again in terms of ἐναντία at b25–6.[13] This is not the language of someone keeping a careful eye on the technical distinctions of *Cat.* 10 or *Metaph.* I. I shall return to this issue at the end of section IV.

It is worth noting the remark at a26–7: 'that everyone in a way makes the principles opposites is, therefore, clear.' Both Simplicius and Philoponus take the πως ('in a way') to qualify 'opposites', and hence, they think, to reflect Aristotle's recognition that full/empty, angular/unangular, straight/curved are not contraries but ἕξεις and privations;[14] Simplicius links this with Aristotle's own appeal to form and privation in his positive account. I do not see why these pairs, suitably understood, should not count as complete differences within a genus; but in any case I think it more natural to take πως to qualify ποιοῦσι, and as thus making the whole assertion slightly guarded.

II. How Should we Understand Aristotle's Various Claims about ἀρχαί at 188a19–30?

This passage divides into two parts. At a19–27 Aristotle invokes the views of some of his predecessors, and at a27–30 he gives an account of why their view that opposites are principles is reasonable.

> Now everyone makes the opposites principles – those who say that the universe is one and unchanging (for even Parmenides makes hot and

[13] Cf. the summing up at the very end of the chapter (189a9–10), discussed below. This usage of ἀντικείμενον is repeated in ch. 7 (190a18 and 26, 190b13; at 190b27 the privation is glossed as 'the opposition' (ἡ ἐναντίωσις)).

[14] Simpl. *In Phys.* 181.30–182.6; Philop. *In Phys.* 117.16–22.

> cold principles, and these he calls fire and earth), or that the principles are rare and dense, or Democritus <who says that they are> the full and empty, of which he says that the one is as what is and the other is as what is not. Further <he says that things differ> by position, shape, arrangement: these are genera of opposites – of position they are above and below, in front and behind; of shape angled and unangular,[15] straight and curved. That everyone in a way makes the principles opposites is, therefore, clear. (188a19–27)

This is, on the face of it, somewhat unexpected: though there has been some talk of opposites, of course, what has preceded in chapter 4 focused at least as much on his predecessors' treatment of 'the underlying thing', and what follows refers only to the Eleatics, the unnamed proponents of the rare and the dense, and the atomists – hardly a complete list. The best explanation is probably that Aristotle, having first deployed a classification in terms of the *number* of principles his predecessors recognised, is here introducing a fresh way of thinking about their view of principles.[16] The incompleteness of his list presumably reflects the fact that he takes other predecessors' appeals to opposites to be obvious.

I am not going to discuss the accuracy of Aristotle's claims about the Presocratics, but I will say something about Aristotle's account of Democritus, and draw a consequence for his account of his other predecessors as well. It does not seem unreasonable to say that Democritus' principles include the full and the empty, and – with a stretch – perhaps the angled and unangular, and the straight and curved as well. It is plain, however, that – despite what Aristotle is going on to say about opposites at 188a30–b26 – these principles are *not*, in Democritus' system, the termini of change: nothing in his system changes from full to empty, or from angled to unangular.[17] In what sense, then, are they principles? It seems that Aristotle is treating his predecessors' appeals to opposites as systems or schemata of fundamental physical explanation, in which the opposites

[15] Why does Aristotle not give pairs of opposites for arrangement (τάξις)? Ross is probably right to suppose that he is being illustrative in his treatment rather than comprehensive; but Aristotle does face a problem over the idea that there are ἐναντία of arrangements: see section V.

[16] I am grateful to Tim Clarke for suggesting this.

[17] Someone might hold an 'atomic' theory according to which the basic entities were not atoms but regions of space which have successive phases of being full and empty – a sort of boolean material field. But this is neither Democritus' view nor how Aristotle understands his view. (Note that some of the opposites cited by Aristotle *can* be termini of Democritean change: an atom can come to be above or behind another.)

figure as the terms in which ultimate physical explanations are to be given. One such schema might take the form:

> *X* comes to be *F* because of the movements of the full in the empty.

But then there seems no reason to rule out explanations which do *not* involve change, and hence schemata of the form:

> *X* is *G* because of the angularity of its atoms.[18]

By the same token, Aristotle may be thinking that, for Anaximenes, at least some ultimate physical explanations take the form:

> *X* is *F* (or even: there is an *F* here) because *X* is (here there is) air, of such-and-such density.

The upshot of this is that the common ground with his predecessors to which Aristotle is appealing is about fundamental explanations in natural science generally, and not exclusively about the explanation of natural *changes*. This brings into sharp relief an issue we would face in any case. In chapter 1 Aristotle says that the causes and principles that he is concerned with are those we need for the science of, or knowledge of, φύσις.[19] Aristotle makes the meaning of this entirely clear in chapter 7: they are 'the causes and principles of the things which are by nature'.[20] The principles Aristotle seeks in *Physics* I are thus not the principles of *change*, as some commentators have held,[21] but the principles of the things

[18] Compare Democritus' explanations of how things taste (Theophr. *De Sens.* 66–7 [= *DK* A135]). On the more general point about the role of principles, compare the complaint Aristotle makes against the atomists at *DC* III.4.303a29–b2 that they do not need (as the atomists suppose they do) an infinite number of elements (στοιχεῖα) because the shapes of the atomic bodies, by which they differ from one another, are composed of rectilinear and non-rectilinear 'pyramids': 'it is necessary for there to be principles of the [atomic] shapes, so that whether these are one or two or more, the simple bodies will be of this number'. This only begins to look like a satisfactory complaint if Aristotle is thinking of the 'elements' as fundamental explanatory principles. (Cf. also, perhaps, *Metaph.* H 2's remarks about the principles of being (1042b31ff.).)

[19] τῆς περὶ φύσεως ἐπιστήμης: 184a10–16.

[20] αἰτίαι καὶ ἀρχαὶ τῶν φύσει ὄντων: 190b17–18. We may compare the similar expressions later in 7.191a3–4 (αἱ ἀρχαὶ τῶν περὶ γένεσιν φυσικῶν; Ross thinks that περὶ γένεσιν is unidiomatic and suggests deleting it; this does not affect my point, which in any case focuses on τῶν φυσικῶν), and at II.1.193a9–10 (ἡ φύσις καὶ ἡ οὐσία τῶν φύσει ὄντων). Further confirmation of the point – if it were needed – is proved by the present passage and by the claim discussed in section III that things 'are from' the principles, in a compositional sense of 'from'.

[21] See, e.g., Bostock 1982: 179–80; the view goes back to Aquinas' commentary on the *De Sensu* (Prohemium 29–34): 'in natural science one proceeds from universals to things which are less universal, as the philosopher teaches in *Physics* I. For this reason he began the teaching of natural science with what is most common to all natural things, namely motion and the principles of motion' (the passage is quoted in Falcon 2012: 527).

which are by nature – and so, principally, of natural substances. One of the principal questions for *Physics* I as a whole is why Aristotle tries to arrive at these principles by investigating *change*. I shall discuss this in section IV.

III. The Tension between the Claims that ἐναντία Are and Are Not from Each Other

> And this is reasonable (εὐλόγως): for the principles must be neither from each other nor from other things, and all things must be from them. These things belong to the primary opposites – on the one hand because primary things are not from other things, and on the other because opposites are not from each other. (188a27–30)

The argument appears to be this:

(1) The principles must not be from each other; and they must not be from other things – the assumption in both cases is that if *X* is from *Y*, *X* cannot be a principle.
(2) All (other) things must be from the principles – the assumption is that principles must be the principles *of* all other things in some way.
Supply: (3) Other things *are* from the primary opposites: clearly Aristotle thinks that he and/or his predecessors are committed to this.
(4) Primary opposites are *not* from other things because they are primary.
(5) Primary opposites are not from each other because opposites are not from each other: it is hard to assess this without investigating the meaning of ἐκ: see below.

Two comments are in order before I turn to the main issue I wish to discuss. The first concerns premise (3). Aristotle neither needs to maintain, nor would be justified in maintaining, that *all* other things come from the primary opposites – unless he wishes to conclude that the *only* principles are these opposites. His premises here are entirely consistent with the claim that there are *other* principles, which do not come from anything. Nor would he appear to be justified were he to hold that his predecessors, at any rate, held the stronger view of (3), given what he has said in chapter 4 about their appeal to the underlying thing. The second concerns premise (4). This might show that the primary opposites are not from other opposites; but the argument does not exclude the possibility that there might be primary things which are not opposites and that the primary opposites are from these. Both of these points are relevant to the arguments of chapters 6 and 7.

The key problem here is that Aristotle appears to insist both that opposites are not from (ἐκ) each other (a27–30) *and* that they come to be from each other (a30–b26). One way to disarm this apparent contradiction would be to suppose that Aristotle is not arguing in his own person at a27–30, and that his claim is only that his predecessors' position is reasonable *from their point of view*. εὐλόγως and εὔλογον in Aristotle's discussions of nature are positive, if more or less cautious, terms: they need not indicate a lack of confidence in the truth of what is in question,[22] though they can certainly indicate that the grounds being offered are not conclusive.[23] Cases in which these terms introduce something which Aristotle takes to be reasonable in a purely *ad hominem* way are hard to find: the closest one is perhaps *GC* I.2.315b30–3:

> This latter doctrine, as we have explained elsewhere, is in itself unreasonable – to halt the analysis at planes. Thus it is more reasonable to hold that what are indivisible are bodies, but a great many unreasonable consequences are involved here too.[24]

Physics III.4.203b4–15 might be more to the point: it introduces a line of thought which makes it reasonable to take the infinite to be a principle, and so it 'justifies' an *endoxon* which Aristotle himself firmly rejects.

There are two problems with reading our passage in this sort of way, however. First, Aristotle is citing his predecessors in support of his own view: this would be problematic if their view were based on a claim which Aristotle rejects outright. Second, his account of Democritus (and perhaps the other Presocratics, too) itself presupposes a sense of 'are from' given which he ought to accept something like the line of thought in a27–30 (see below).[25]

[22] See, e.g., *Phys.* III.7.207b1–5, IV.12.220b24–6, VIII.5.256b13–24, VIII.9.265a27–b16; *GC* I.7.324a9–11, II.3.330b1–7, II.10.338a14–b1; *PA* II.1.647b4–6; *Metaph.* Λ 8.1074a24–31. An interesting example is the remark at *GA* I.23.731a24: καὶ ταῦτα πάντα εὐλόγως ἡ φύσις δημιουργεῖ. (Another usage common in the biological works – perhaps not unrelated to this idea – is to introduce a fresh set of considerations to show that something already established might have been expected in any case: see, e.g., *PA* II.2.648b8, II.4.651a12–13, II.6.652a7–8, III.4.666b13–14; *GA* I.1.715b8, II.4.738a16–22.)

[23] See, e.g., *HA* VI.13.567b9–11; *GA* I.1.715b13, II.7.747a3–4; *Metaph.* Λ 8.1074a16–17. They are also Aristotle's preferred terms when dealing with cosmological matters when proof and/or direct confirmation are not in the offing. Instructive examples include *DC* II.2.284b18–24, II.8.290a1–5, IV.4.312a5–8; *GC* II.10.336b25–34; *Meteor.* I.3.341a23–6.

[24] τοῦτο μὲν οὖν αὐτό, καθάπερ καὶ ἐν ἄλλοις εἰρήκαμεν, ἄλογον μέχρι ἐπιπέδων διαλῦσαι· διὸ μᾶλλον εὔλογον σώματα εἶναι ἀδιαίρετα, ἀλλὰ καὶ ταῦτα πολλὴν ἔχει ἀλογίαν (translation from Williams 1982). *Meteor.* II.2.354b4–10 is, in a way, another case: δόξειε γὰρ ἂν εὔλογον εἶναι, καθάπερ καὶ τῶν ἄλλων στοιχείων κτλ; but there εὔλογον is explicitly within the scope of δόξειε.

[25] For these reasons, while Aristotle may take his predecessors' use of 'is from' to be somewhat vague, I think that he must have a determinate usage in mind here which yields a claim which he can

If Aristotle is endorsing, or at least not rejecting, the premises of the argument, then there are, in principle, at least two ways of dealing with the problem: we might suppose that the claims invoke opposites in two different ways – for example, by referring to opposite properties, on the one hand, and to subjects qualified by those properties, on the other – or that Aristotle is using 'from' in 'are from' and 'come from' in two different senses. The famous discussion of opposites in the *Phaedo*,[26] which no doubt Aristotle has in mind here, might suggest the first of these. Socrates says (A) that every opposite necessarily comes to be from its opposite, and (B) that opposites do not come to be their opposite.[27] An unnamed interlocutor says that (B) 'is the very opposite'[28] of (A): Socrates resolves the problem by explaining that (A) was a claim about 'the thing that is opposite',[29] whereas (B) is about 'the opposite itself'.[30] Is Socrates' resolution Aristotle's too? The first thing to notice is that in affirming (B) Socrates does not actually say that opposites do not *come from* their opposites: he says that they are not willing to *come to be* the other opposite, but rather withdraw or perish at its approach (102d5–103a2); nor does the unnamed interlocutor rephrase (B) in this direction. So Socrates' problem is not obviously the same as Aristotle's problem. Might the solution nonetheless be the same? It is striking that (as we shall see in section IV) Aristotle shows little or no interest in making clear the ontological status of the opposites as they appear in either of the 'ἐκ' claims, and this alone might be enough to entitle us to suppose that he is not mirroring the *Phaedo*. Let us suppose, however, that this is his solution nonetheless.

(by his lights) reasonably ascribe to his predecessors, and also agree to himself (albeit, perhaps, in a refined or modified way). (If he does accept the line of thought at a27–30, why does it not itself constitute an argument ἐπὶ τοῦ λόγου? Perhaps because Aristotle thinks that one should begin the enquiry into principles by looking, not for the constituents of things, but at what makes them able to change. See further pp. 140–41.)

26 70d7–71a10, 102d5ff.

27 (A) τοῦτο οὖν σκεψώμεθα, ἆρα ἀναγκαῖον ὅσοις ἔστι τι ἐναντίον, μηδαμόθεν ἄλλοθεν αὐτὸ γίγνεσθαι ἢ ἐκ τοῦ αὐτῷ ἐναντίου (70e4–6); (B) οὐδ' ἄλλο οὐδὲν τῶν ἐναντίων, ἔτι ὂν ὅπερ ἦν, ἅμα τοὐναντίον γίγνεσθαί τε καὶ εἶναι (102e7–103a1).

28 οὐκ ἐν τοῖς πρόσθεν ἡμῖν λόγοις αὐτὸ τὸ ἐναντίον τῶν νυνὶ λεγομένων ὡμολογεῖτο, ἐκ τοῦ ἐλάττονος τὸ μεῖζον γίγνεσθαι καὶ ἐκ τοῦ μείζονος τὸ ἔλαττον, καὶ ἀτεχνῶς αὕτη εἶναι ἡ γένεσις τοῖς ἐναντίοις, ἐκ τῶν ἐναντίων; (103a5–9).

29 τὸ ἐναντίον πρᾶγμα (103b2–4). Cf. 71b2; more generally, the earlier passage talks in terms of *something coming to be F.*

30 αὐτὸ τὸ ἐναντίον (103b4–5). Note that this may not quite be the same as the distinction between a property and the object that has the property, since Socrates will later go on to talk indifferently of fire and threeness as being what brings the opposite to the object that has it. Equally, the notion of the opposite itself is unclear: the talk of its withdrawing and ceasing to be suggests that it is not an abstract item (e.g. a pure universal), nor the particular instantiation of the property in a given object, since its 'withdrawal' is distinguished from its ceasing to be.

The claims would be that a pair of opposites (say, *F* and *G*) are themselves not from each other, whereas a thing which is *F* comes from something which is *G*: these are at least something like the claims that *F itself* (perhaps the property *F*) does not become *G*, but things which are *F* do become *G*. While this would be compatible with claim (1) of the argument at a27–30 – that principles are not from each other – it would make a nonsense of claim (2), that things which are not principles are from the principles. So while the ontological status of 'the *F*' may be an important issue in its own right, I think the most promising solution to our problem is to suppose that Aristotle is using 'are from' and 'come to be from' in two different senses. This will involve a sense of ἐκ in which opposites plainly are not from each other – but at the same time other things are from them – and another sense in which opposites do come from each other. As with his account of Democritus, discussed earlier, this again is hard to square with the idea that Aristotle is all along talking about the principles of *change* or about opposites solely as the termini of changes.

Aristotle's account of the senses of ἐκ in *Metaph.* Δ 24 is the natural place to look; but this account is not altogether helpful, and is in any case far from perspicuous. He distinguishes the following ways in which something can be said to be that from which something is:

- (i) as matter (in some sense: but does Aristotle mean the matter that now underlies the thing, or the preceding item from which it came to be?)
- (ii) as efficient cause
- (iii) as a whole which a part is from
- (iv) as the part a form is from
- (v) as a thing something is from a part of
- (vi) as what precedes something in time, whether merely preceding it or being replaced by it.

Of these, only (i) and possibly (vi) seem relevant; strikingly, one sense that seems lacking is the sense in which a thing is from its components – the converse, as it were, of (iii) – indeed, one type of case of this (but not the only one) is mentioned in Aristotle's formulation of (iii): 'what is composed from matter and form'.[31] Perhaps Aristotle intends this to be

[31] ἐκ τοῦ συνθέτου ἐκ τῆς ὕλης καὶ τῆς μορφῆς (1023a31–2). We find ἐκ used in a clearly compositional sense at *GC* I.2 316b4–5 (ἄτοπον ἐκ μὴ μεγεθῶν μέγεθος εἶναι) and *Phys.* IV.10.218a7–8 (συγκεῖσθαι δεῖ τὸ ὅλον ἐκ τῶν μερῶν· ὁ δὲ χρόνος οὐ δοκεῖ συγκεῖσθαι ἐκ τῶν νῦν) – and in relation to the Presocratics' view at *Metaph.* Δ 2.1004b29–30: τὰ δ' ὄντα καὶ τὴν

covered by (i), but if so this is hardly clear. In any case, the compositional sense seems a plausible one to see deployed at a27–30: opposites are not composed of each other, while principles (sc. in the relevant sense) are what other things are composed of. Now 'composed of' is not itself a transparent notion. We might distinguish – at least – the following:

(A) *X* is composed of *Y*, *Z*, . . . iff *Y*, *Z*, . . . are its non-overlapping proper physical parts
(B) *X* is composed of *Y*, *Z*, . . . iff *Y*, *Z*, . . . are its metaphysical parts (e.g. its form and matter)
(C) *X* is composed of *Y*, *Z*, . . . iff *Y*, *Z*, . . . are the matter for its physical parts
(D) *X* is composed of *Y*, *Z*, . . . iff *Y*, *Z*, . . . are the features of its physical (or metaphysical) parts, or of the matter of its physical parts, which differentiate *X* from other things
(D*) *X* is composed of *Y*, *Z*, . . . iff *Y*, *Z*, . . . are the terms in which the features of its physical (or metaphysical) parts, or of the matter of its physical parts, which differentiate *X* from other things, are essentially characterised.

With (D) and (D*) we come close to the thought that seems to lie behind Aristotle's treatment of Democritus – or indeed of those who make the underlying thing (say) air, and for whom the principles are the rare and the dense.[32] On these readings, at least, the claims that opposites are not composed of each other, that principles are not composed of other things and that other things are composed of principles seem quite plausible. With this usage – and this idea – we may compare *Metaph.* I 7.1057b22–3: 'the opposites are not composed of each other, and so are principles'.[33] The sense of 'come to be ἐκ' in play at a30–b26 will indeed turn out to be different from this 'compositional' sense.[34] There is, of course, a further

οὐσίαν ὁμολογοῦσιν ἐξ ἐναντίων σχεδὸν ἅπαντες συγκεῖσθαι (Diana Quarantotto drew my attention to this passage).

32 (D) or (D*) also fit Aristotle's treatment of Parmenides better than (A) or (C): the hot and the cold are more plausibly features of fire and earth than they are the referents of the terms 'fire' and 'earth'.

33 τὰ μὲν οὖν ἐναντία ἀσύνθετα ἐξ ἀλλήλων, ὥστε ἀρχαί (the contrast there is with the intermediates, which σύγκειται ἐκ τῶν ἐναντίων (1057b33–4): see p. 147 below).

34 This is close to part of Simplicius' view (Simpl. *In Phys.* 182.19–30). Like Philoponus (Philop. *In Phys.* 111.17–26) he takes it that opposites do not come from each other in the sense that neither opposite can be the (pre-existing and persisting) matter for the other's coming to be, nor its efficient cause (186.20–1); but he takes the claim that opposites are principles to be true because 'the coming to be of things which come to be after the principles involves change; and every change, as will be shown, comes about from *enantion* to *enantion*' (182.28–30). For his reading of the claim that opposites do come from each other (186.15ff.), see section V below. (According to Simplicius,

question which arises in chapter 6 – whether when Aristotle implies there that a substance 'is not from' non-substances (πῶς οὖν ἐκ μὴ οὐσιῶν οὐσία ἂν εἴη; (189a33)), he is using 'is ἐκ' in the same sense as the sense identified here.[35] It is not obvious that substances cannot be from non-substances in sense (D) or sense (D*) – whereas it might be closer to being obvious if we were talking about sense (B).

If this passage is concerned with composition and not with the termini of change, what is the connection between the enquiry into principles and the investigation of changes? As I said above (p. 136), this is one of the central questions for *Physics* I as a whole. One possibility, quite in line with the general approach of chapter 5, is that Aristotle thinks that if all changes are between opposites, any natural substance must be stocked, as it were, with (members of pairs of) opposites: any feature a substance had which is not an opposite, or which is not suitably underpinned by opposites in the way exemplified by the σύνθετα at b8ff.,[36] would be a feature in respect of which it simply could not change. This line of thought does not obviously get us to the *principles* of natural things, however. My own view is that in *Physics* I Aristotle gives us only a partial glimpse of his overall strategy – as, indeed, he only gives us a glimpse of the notions of matter and substantial form.[37] While there simply is no deducing the principles of natural things from a general analysis of changes, what Aristotle will do elsewhere is to develop an understanding of matter as what underlies substantial generation and destruction, of what form must be like in these cases, and of the roles of potentiality and actuality, in such a way as to make it very plausible that the principles of natural substances are indeed the fundamental items involved in change, namely matter and form.[38]

IV. The Logical Grammar and/or the Ontological Commitments of Aristotle's Claims about ἐναντία at 188a30–b26

> We must, however, investigate how this turns out on the basis of general considerations. (188a30–1)

Ammonius agreed that Aristotle uses ἐκ in two different senses, but offered a Platonising interpretation of the first sense: principles are not ἐκ other things because they are the highest genera, and hence prior to everything else (183.18–35).)

35 That the senses are the same is taken for granted by Kelsey (2008: 186–8).

36 See section IV.

37 And, one might add, virtually no sight at all of the notion of potentiality.

38 I discuss this further in Judson (forthcoming), Prologue to ch. 2. David Charles argues for a related but distinct position in his chapter on I.7.

We should probably associate the phrase ἐπὶ τοῦ λόγου – 'on the basis of general considerations' – with the more familiar notion of arguments which proceed λογικῶς.[39] Aristotle frequently contrasts such arguments with those that proceed in some other way. In the *De caelo* the contrast is between relatively general arguments and ones that proceed φυσικῶς – that is, which appeal to specific properties of the cosmos and/or of the natural bodies concerned. In *Metaph.* Z the other side of the contrast is left implicit: what seems to be involved is the idea of being more neutral or less neutral between competing metaphysical views.[40] ἐπὶ τοῦ λόγου itself is also used at *Phys.* VIII.8.262a18–19 and *GA* II.4.740a4–5, where the contrast is between basing a conclusion on observation and basing it on a more general argument. The appeal to his predecessors' views *is* an argument, Aristotle thinks; but now he gives a more general one. Some commentators take the claim that the schema set out at 188a36–b26 applies to all cases of change to be meant as a simple conceptual truth reflecting the very nature of change: there is just no *change* if *S* comes to be *A* and *A* is not an opposite/intermediate or if *S* (or something appropriately related to *S*) did not possess *A*'s opposite/intermediate earlier.[41] Aristotle does not appear to endorse the claim understood this way (which would commit him to thinking that something's coming to be a house is not a *real* change, and that the real change is something coming to be house-structured), as his treatment of the cases of the house and the statue seems to show. He is right not to endorse the claim, as it is simply false: it is enough, at this conceptual level, for a real change that *S* or its appropriately related predecessor failed to be *A*.[42] Given this, and the context of Aristotle's discussion, it seems better to suppose that he intends his claim as a deeper scientific

39 Charlton floats the idea that this means 'with reference to how we speak', and cites *Phaedo* 99e4–100a2 in support of this. But λόγοι in that passage does not obviously refer to speech or how we speak rather than to arguments or accounts. (ἐπὶ τοῦ λόγου does mean something like this (with the idea of being taken in by how one expresses something) in a number of passages in the *SE*.) Note that Charlton misrepresents Ross, who associates ἐπὶ τοῦ λόγου in I.5 with the use of the phrase at *Phys.* VIII.8.262a18–19 mentioned above, and not, as Charlton says, with *GC* I.8.325a14.

40 For discussion see Burnyeat 2001, esp. 19–25; Lewis 2013: Introduction and ch. 1.

41 This may be Broadie's view – or at least her view of Aristotle's view of the Presocratics. At any rate, she speaks, in the latter connection, of the 'conceptual advantages in equating X's coming into being with X's coming into being from the opposite of X, say A' (Waterlow 1982: 10). For Kelsey's view see n. 47. Charlton's view is that it 'is not an empirical doctrine to the effect that the universe is regular; it is the purely logical doctrine that change is within definite ranges' (1970: 66). I am not sure that I understand the force of 'purely logical', but in any case, the 'doctrine' seems either simply false or too weak to capture Aristotle's claim about the role of ἐναντία.

42 In saying this I am simply bracketing the need for a further condition so as to exclude Cambridge changes: this is a separate issue which arises for both the views of change in question here.

truth, applying (as we shall see) either to changes or to underlying changes as appropriate: general reflection on these cases reveals an articulated explanatory pattern which his predecessors did not (or did not properly) discern.[43]

Before turning to the claims at a31ff., we should ask what the contrast is between simples (ἁπλᾶ) and composites (σύνθετα) at b8–10:

> And this holds in the same way in the other cases, since the same account holds also for those of the things that are which are not simple but composite. (188b8–10)

We might expect the composites to be things composed of simples – so that the composites would be composed of a plurality of opposites. Aristotle does not always use 'composite' in this way, however.[44] More importantly, if the composites here were thought of in this way, their coming to be ought to be analysed in terms of the joint coming to be of their component opposites; but it is in fact analysed in terms of distinct and apparently unitary opposites.[45] So I suggest that composites here are composites in a more general sense, namely things whose coming to be is a matter of the combining or arranging of a number of things or parts, and that the 'simples' are just items which are not composites in this sense.[46]

Aristotle's principal claim in this passage appears to be 'of all the things that are nothing is of a nature to do or to undergo just anything under the agency of just anything; nor does anything whatever come to be from anything whatever.' It is the latter claim on which he expands at some length; as we shall see (pp. 146–7), this will in turn justify the former claim. This elaboration takes the form of a series of examples which plainly stands for, or

[43] An example of this type of interpretation is that of Bogen, who thinks that Aristotle's claim invokes *abilities to change*: 'being dark seems to involve more than privation of lightness; to be dark is also to have a predicate which makes a subject able to become light' (1991: 60; cf. Simpl. *In Phys.* 186.15–187.9). For further discussion see pp. 140–41 above and Charles's chapter on I.7. Robert Bolton claims that Aristotle's appeal to different cases of change is a straightforward case of Aristotelian induction or 'empirical reasoning' (1991: 23–6). I think that this seriously underplays both the characterisation of the argument as ἐπὶ τοῦ λόγου (see above) and Aristotle's general argumentative strategy in *Phys.* I: Aristotle thinks that his predecessors have latched onto the truth up to a point, but have not been able to give an account (188b26–30). To use the language of ch. 1, they have not progressed (far) beyond the level of what is compounded (συγκεχυμένα), and what is needed to progress to what is more intelligible by nature is not observation, but rather rational reflection on what we observe. I am grateful to Alan Code for pressing me on this point.

[44] So, e.g., the composites of *Metaph.* Z 4, such as (a) pale man, are contrasted not with their components as such, but with genuine unities.

[45] Or at least things which Aristotle claims here to be opposites: see below.

[46] Note that this would not prevent these simples from being composites in the Z 4 sense – e.g. (an) educated human. I discuss the ontology of the simples below.

encourages us to envisage, a schema which is supposed to apply to all cases of change. The claim 'nor does anything whatever come to be from anything whatever' is grounded in the claim that the values for *X* and *Z* in the schema must invoke opposites, and the values for *W* must invoke intermediates between opposites (in cases where there are such intermediates):[47]

ἁπλᾶ	*X* comes to be	not from just any *Y*	but rather from *Z*	or from *W* (if any)
	λευκόν μουσικόν	οὐ λευκοῦ οὐ μουσικοῦ	μέλανος ἀμούσου	τῶν μεταξύ
	X ceases to be	not into just any *Y*	but rather into *Z*	or into *W* (if any)
	τὸ λευκόν τὸ μουσικόν	τὸ μὴ λευκόν τὸ μὴ μουσικόν	τὸ μέλαν τὸ ἄμουσον	τὸ μεταξύ
σύνθετα	*X* comes to be	not from just any *Y*	but rather from *Z*	
	τὸ ἡρμοσμένον τὸ ἀνάρμοστον		ἀναρμόστου ἡρμοσμένου	
	οἰκία		ἐκ τοῦ μὴ συγκεῖσθαι ἀλλὰ διῃρῆσθαι ταδὶ ὡδί	
	ὁ ἀνδριὰς καὶ τῶν ἐσχηματισμένων τι		ἀσχημοσύνης	
	X ceases to be	not into just any *Y*	but rather into *Z*	
	τὸ ἡρμοσμένον		not just any ἀναρμοστίαν, but τὴν ἀντικειμένην	

[47] Cf. b21–6. Kelsey takes quite a different view – that the 'not just anything' claim is the ground for the 'opposites/intermediates' claim. He then identifies a gap between them, which he thinks can be filled only if we suppose that 'Aristotle is operating in *Physics* I.5 with a very specific and (we might say) one-sided conception of change, according to which the action of change upon its subjects is essentially destructive' (2008: 188–90; this quotation p. 190). He detects this conception of change in the idea of opposites 'driving themselves out of their own nature' at *GC* I.7.323b25–9. But if there is a gap, it seems to me simply to recur in the same form in the 'driving out' claim: we might grant that not just anything whatever can drive just anything out, but that falls short in exactly the same way as the idea that whenever something is driven out it must be a case of an opposite driving an opposite out.

This passage raises a number of questions.

(i) The meaning of 'come to be from'. Aristotle makes it clear that this is to be understood as parallel to 'cease to be into'. The suitability of *both* expressions already tells us that 'from' here does not have a compositional sense, and the parallel with ceasing to be into makes it clear that what is involved is temporal succession. We met a sense of ἐκ like this in *Metaph.* Δ 24 (sense (vi)): there Aristotle distinguishes what we might call *mere* temporal succession from succession which involves *replacement*: it will turn out that he needs a yet stronger sense here, in which the replaced and replacing items are identical or related in some special way. I shall return to this point.

(ii) The ontological status of the values for *X*, *Y*, *Z* and *W* – and hence the nature of the 'invoke' relation. As is clear from the table above, Aristotle uses an almost perversely varied set of expressions for the values for *X*, *Y*, *Z* and *W*: neuter singular adjectives without an article, neuter singular adjectives with a definite article, abstract nouns – and for good measure concrete nouns as well. Expressions such as τὸ λευκόν could refer to the property *being pale*, or to a thing which has the property – the pale thing. The use of abstract nouns such as ἀσχημοσύνης (and perhaps the use of the adjective without an article) might lead us to suppose that the former usage is in view, as might the phrase τὰς ἀντικειμένας διαθέσεις at b10–11;[48] but the latter usage is suggested by the expression τοῦ μὴ συγκεῖσθαι ἀλλὰ διῃρῆσθαι ταδὶ ὡδί at b18–19,[49] and by Aristotle's glossing ἀνδριάς as an example of τῶν ἐσχηματισμένων. Bogen points to the expression πλὴν εἰ μὴ συμβεβηκὸς εἴη τῷ μὴ λευκῷ ἢ τῷ μέλανι τὸ μουσικόν at a35–6, and argues that 'the dative construction of the expression ... τῷ μὴ λευκῷ makes it look as though the expression "the musical" must refer (not to a subject which is musical, but rather) to something ... which can be in a subject'.[50] This argument seems a two-edged sword, however, since if it were right the subsequent expressions τῷ μὴ λευκῷ and τῷ μέλανι would have to refer to a *subject* which is not-pale or dark. In any case, I prefer to understand συμβεβηκός as used as a nominal rather than an adverbial expression, and to translate this as 'unless the educated were an incidental to the not pale or the dark', which carries neither implication.[51]

48 Cf. Charlton 1970: 70. For Aristotle's account of διαθέσεις as qualities see *Cat.* 8.8b25–9a13.

49 *Pace* Charlton ibid.

50 Bogen 1991: 60. Bogen does not say how he would translate the expression, but I suppose that he must be taking συμβεβηκός to be elliptical for κατὰ συμβεβηκός, and that he would render it 'unless the musical is in the not pale or the dark incidentally'.

51 Cf. III.4.203a5 (οὐχ ὡς συμβεβηκός τινι ἑτέρῳ ἀλλ' οὐσίαν) and II.5.196b35 (συνέβη αὐτῷ ἐλθεῖν). Note that the parallel expression for the case of ceasing to be, a few lines later in I.5 (188b4–5), has, on the one hand, κατὰ συμβεβηκός, but, on the other hand, no datives. Charlton seems to construe

However we are to construe the values of *X*, *Y*, *Z* and *W*, it seems plausible that Aristotle must take the ultimate truth-makers of these claims to be something's coming to be *F* from something's being *G* (where *F* and *G* are placeholders for opposites or intermediate properties). Suppose that this passage envisaged the weaker condition that:

(C1) If *F*(ness) comes to be at t_2, then there was *G*(ness) at t_1[52]

As we saw, Aristotle's principal claim in this passage as a whole is 'of all the things that are nothing is of a nature to do or to undergo just anything under the agency of just anything; nor does anything whatever come to be from anything whatever.' Although statements of the form of (C1) may state a necessary condition for this claim, it is hard to see how they could establish either of its parts. If the requirement for *x* coming to be *F* is simply that there was *G*(ness) earlier, then it would seem that there are no constraints on what *x* must be like earlier, and hence that anything might come to be from anything whatever, and equally that an agent could cause something to undergo anything – it would only be that nothing could come from anything *in any circumstances whatever*. Indeed, for the same reason, even the condition

(C2) $\forall x$ if *x* comes to be *F* at t_2, then $\exists y$ *y* was *G* at t_1

is too weak: we need one or other of:

(C3) $\forall x$ if *x* comes to be *F* at t_2, then $\exists y$ R*xy* and *y* was *G* at t_1

(C4) $\forall x$ if *x* comes to be *F* at t_2, then *x* was *G* at t_1

where R is some relation which would distinguish cases of genuine change between opposites from other cases for which only (C2) might be true; this relation might be disjunctive, with one disjunct being identity. The question whether Aristotle has (C3) or (C4) in mind is of course the question whether he thinks that the subject of a change persists through every case of change – and that is raised, if at all, only later in book I. If the ultimate truth-makers of the claims encapsulated in the schema are of the form of either (C3) or (C4), we can see how Aristotle supposes his

the expression in a similar way to me: 'unless the knowing music supervenes on the not pale or the dark'. Hardie and Gaye (1930) side with Bogen: 'unless "musical" happened to be an attribute of the not-white or of the black'.

52 For ease of exposition (C1)–(C4) here do not attempt to capture the further key feature of Aristotle's claims, that whenever *any* change takes place, it is or involves a change between opposites and/or intermediates.

principal claim to be justified. Nothing can come to be *F* without it or something appropriately related to it being *G* earlier; and so nothing comes to be from just anything whatever; nor will just any agency suffice, for it must be one that can act on what was *G* so as to make it or its appropriately related successor *F*.[53]

(iii) What is the distinction between the permissible values for *Y* (οὐκ-*F*) and those for *Z* (privative-α-*F*)? Despite my critical comments on the use of the term 'contraries' for what is invoked by the intended values for *X* and *Z*, I think that it is quite reasonable to think of what is invoked by the values for *Y* in terms of *contradictories*: Aristotle regularly treats being not-*F* as a matter of simply failing to be *F*.[54] One way of *X*'s failing to be *F*, for Aristotle, is for *X* not to exist;[55] another is for *X* not to be the sort of thing which can be *F*. So one might think that the contrast between privative-α-*F* and οὐκ-*F* is primarily that between things which fail to be *F* but which are capable of being *F* and things which are simply not capable of being *F*. This does not, however, capture the whole of Aristotle's notion of ἐναντίον, since one other element in this idea, as we have seen, is that of maximum difference within a genus – this is, for instance, why intermediates are not simply further ἐναντία – and more generally, there is his insistence that each ἐναντίον only has one opposite. Another element in the idea is that Aristotle often regards *both* opposites as being or involving causal powers.[56] Even when he characterises contraries in terms of ἕξις and privation in *Metaphysics* Ι,[57] he says that intermediates are 'composed of' the two contraries:[58] this difficult idea only begins to make sense if we think of the intermediates as a blend of two causally efficacious ingredients, rather than as a blend of a form and its absence. On the other hand, later in *Physics* I Aristotle seems willing to treat privation as simply the absence of the form: see 7.191a5–7 and 9.192a3–5.[59]

[53] Simplicius thinks that Aristotle is basing what I have called his principal claim here on a causal transmission principle – that contraries are needed to *act* to produce their like (Simpl. *In Phys.* 184.14–186.14). As Simplicius admits at 185.34–6, however, none of the examples in Aristotle's schema are cases of agency, so this interpretation is very hard to sustain.

[54] See, e.g., *Int.* 10.

[55] *Cat.* 10.13b15–33. Aristotle is – apparently – not entirely consistent in this view, since he thinks Homer is a poet (*Int.* 11.21a25–6): cf. Castelli 2018: *ad* 1057a30–b1. For a defence of Aristotle see Mignucci 2007: 125–30.

[56] Hot and cold are both active: *Meteor.* IV.1.378b10–379a3 (cf. 12.390b2–9), *GC* II.2.329b24–30; dry and moist, though passive, both have causal characteristics: *GC* II.2.329b30–330a12. Cf. Bogen 1991: 60.

[57] See n. 12 above.

[58] *Metaph.* Ι 7.1057b22–34; cf. p. 140 and n. 33 above.

[59] On the other hand again, see *Phys.* II.1.193b18–21.

(iv) The οἰκία/ἀνδριάς cases introduce two new points. First, Aristotle introduces (admittedly only by implication) the idea that the claim that change is between opposites may in some cases only apply to *underlying changes*: Aristotle is not saying that οἰκία has an opposite, nor that there is anything scientifically improper or misleading about saying that something comes to be a house or that a house comes to be. Rather, he seems to think that such apparently opposite-less changes occur in virtue of some underlying change which is between opposites.[60] Second, Aristotle applies the locution 'not just any ... but the ... ', which served to pick out the opposite from other kinds of 'contradictory' to the (alleged) opposite itself:

> For it is necessary for everything structured to come to be from unstructured, and the unstructured from structured, and for the structured to cease to be into unstructuredness – and this is not just any unstructuredness but the opposed one. (188b12–15)

Presumably this is based on the recognition that a set of building materials perfectly capable of, and suitable for, being built into a house can instantiate many conditions which are instances of lacking house-structure but which are structurally and causally distinct from each other and which cannot be regarded as intermediates on a single spectrum: it could be structured so as to form a temple, a council-chamber, a bridge, etc. None of these conditions is the mere 'contradictory' property of failing to be structured as a house. Aristotle now seems tempted by the idea that 'maximum difference' should nonetheless come into play to single out one of these conditions as the furthest from being structured as a house. He betrays some awareness of the shift that has occurred by replacing ἐναντίον by ἀντικειμένην – though, as we saw in section I, his ultimate conclusion will be expressed in terms of ἐναντία once again. But in any case what could this maximally different condition be? Perhaps the obvious – the only? – candidate would be the materials lacking *any* structure: their being just a heap. If so, however, these cases will lose another feature of standard opposites as Aristotle characterises them, namely that each opposite has only one opposite: for this same lacking of *any* structure

60 Ross (1936: 489) takes a different view. He thinks that Aristotle holds that οἰκία, ἀνδριάς, etc., are (potentially misleading) *names* for those very conditions of being structured which are the opposites of the relevant unstructured condition(s). This is certainly a possible way to understand the remark that 'this is overlooked because the opposed conditions have not been named', but it is impossible to square with Aristotle's standard view that substances have no opposite (*Cat.* 5.3b24–32; *Phys.* I.6.189a27–34) – and indeed with the view that a house is something that possesses or exemplifies a certain structured condition. That said, if we look ahead to the doctrines of *Metaph.* Z and Λ (and perhaps even to *Phys.* I.7), Aristotle will not be able to maintain all three of the following: (i) form is a substance; (ii) form is the opposite of privation; (iii) no substance has an opposite. Proposition (ii) is clearly the best one to be dropped. See also section V.

would be maximally different from every condition of being structured as a house, bridge, temple, etc.

One of the most puzzling features of *Physics* I – and equally of its reprise in *Metaph.* Λ 1–2 1069b3–14 – is Aristotle's insistence on retaining the term ἐναντία for the termini of change; even though, as we have seen, he occasionally reaches for the more appropriate term ἀντικείμενα, he nonetheless reverts to ἐναντία; and in the *Metaph.* Λ passage he goes out of his way to say that ἀντικείμενα is too general a term, and that ἐναντία is the precise one.[61] We might have expected Aristotle to say 'the Presocratics and Plato were right, up to a point, to offer fundamental explanations of natural changes and other natural things in terms of items which are ἐναντία, such as the hot and the cold. What they failed to see was, first, that these explanations are too specific, and that there is a general pattern to which the termini of all change, and indeed all scientific explanations of natural substances, conform; and, second, that the most general form of this pattern is not ἐναντίον/ἐναντίον, but form/formless; and the variety of cases are held together by analogy.' I have no good explanation for why he does not say this: the best that I can suggest is this. Perhaps Aristotle finds the appeal to opposites at the fundamental level in the natural philosophy of his predecessors (as he understands them) to be *so* pervasive that he cannot bring himself to suppose that this appeal is utterly wrong: he would rather make (or try to make) his own distinction of form and the formless conform to this universal pattern than to suppose that what everyone thinks is simply false.

V. Assessment of Aristotle's Position in 188a30–b26

Aristotle's focus on opposites creates severe problems for his claim that his schema applies to all cases of change.

(i) Things can come to be dark from being colourless as well as from being pale,[62] and similarly bitter from being flavourless, etc. The same applies, of course, to ceasing to be. This suggests that, if the idea is that things come to be *F* only if they have an ability or potentiality to be *F*, this potentiality should if anything be associated with the thing's particular way of lacking *F* (or, in this sense, its privation of *F*), and not with its possession of the opposite of *F*.[63]

[61] See, on the other hand, *Phys.* V.1.224b28–9, cited on p. 151 below.

[62] See Bostock 1982: 190.

[63] For what I take to be similar reasons Simplicius tries on Aristotle's behalf to align the notions of opposites and privations in the passage cited in n. 43 above.

(ii) Aristotle's claim at *Phys.* V.2.226a23–32 that 'in each of these [change with respect to quality, quantity and place] there is opposition (ἐναντίωσις)' seems to be broadly false for changes in quantity and place.[64] Such changes do involve passing through intermediate points or stages on a continuum, and Bogen and McGuire claim that this is enough to secure the parallel with qualitative opposites.[65] But these continua do not, in general, have extremes or end points other than those defined by the nature of particular changes, and hence they neither constitute maximum differences nor do they come in unique pairings – and occupying one point on one of these continua has, in general, nothing to do with the thing's ability to move to or occupy another point other than providing the merely 'contradictory' property of being somewhere else.[66] In these cases even the modified view I outlined in the previous paragraph – that the potentiality to be *F* should be associated with the thing's particular way of lacking *F* – seems hard to sustain.

(iii) As we have seen, it is hard to regard form and formlessness, or even form and the '*opposed*' formlessness, as opposites. Imagine a heap of sixteen cubic bricks: these can be arranged (e.g.) as a 4 x 4 square, a 2 x 8 rectangle, or a 1 x 16 rectangle. Even if we grant that among the various ways of the bricks' lacking an arrangement there is one which is 'set against' being arranged as a square, it is hard to see how this can be distinct from the one set against being arranged as a 2 x 8 rectangle. But if that is right, these opposites are not uniquely paired with each other; by the same token, the idea of maximal difference within a genus or a single range of change seems to have been lost. These points are related to the fact that the other states of being formed (e.g. the rectangles) are not intermediate states between complete formlessness and being formed as a square.[67] Parallel considerations would apply to a block of marble and being a statue of a particular form. The idea of maximum difference within a genus is perhaps easier to sustain in the case of the coming to be of natural organisms: one extreme is the formed state of the fully developed adult, and the other (on Aristotle's account) is the 'formless' but nonetheless structured state of the material provided by the mother. Chance mutations aside, the latter is only the starting point for

[64] Indeed, in *Cat.* 6 (5b11ff.) Aristotle denies that quantities (or, apparently, places) have ἐναντία. For discussion see Ackrill 1963 ad loc. and Gregoric 2006.

[65] Bogen and McGuire 2006: 5–6 and n. 6.

[66] There may of course be limits to growth given by the nature of the subject if it is an organism (a point made by Bogen (1992: 20); but this need not be true for all changes in quantity in such subjects (such as what we call 'growths'), and is not true for other subjects such as ice-sheets or spits of sand deposited by the sea.

[67] Cf. Philop. *In Phys.* 113.9–114.16, and Bostock 1982: 190.

the development of organisms of one kind, and mature organisms of other kinds come to be from different 'formless but nonetheless structured' maternal states.[68] It has to be said, however, that this neat scheme breaks down in what Aristotle claims to be the corresponding case of ceasing to be. (a) There simply is no single extreme formless but nonetheless structured maximal end-state of ceasing to be a mature adult of a given kind which is unique to each kind of organism. We might try to stipulate that this is the first state in which the organism is dead. Even if we ignore deaths by violence and misadventure, however, bodies at this point will display a great variety of conditions with respect to the characteristic functions of the living organism: some organisms will die while most of these functions were still fully operative, but others will have had (e.g.) hearing, sight or mobility lost or seriously degraded much earlier (or lost limbs or parts of limbs), and hence will be at death 'further away' from the state of the mature organism. All that these putative end-states will have in common is 'being the state of the body of an organism of the kind just after its death', and for the reason given this does not seem to constitute a 'maximum' difference – indeed, it does not seem to constitute a *single* condition at all. (b) Even if there were such a single state for each kind of organism, it manifestly would not be the same condition as that from which the organism came to be – nor would either of these be an intermediate. So the 'structured' condition of the mature organism would have two opposites.[69]

(iv) A related point starts from Aristotle's remark at 188b3: 'nor, then, does anything cease to be into just anything first.' As a concession that *subsequently* the thing might decay further, this might seem judicious, at least in the case of composites. Aristotle cannot afford to make this concession, however: if *X* has ceased to be *F* and has become *G*, there is nowhere for it to go, so to speak, but back to *F* (or to some intermediate) again.

Aristotle seems to acknowledge at least some of these difficulties at *Phys.* V.1.224b28–9, where he says that (non-incidental) change is between opposites, intermediates *and contradictories*;[70] the passage in the following chapter, *Phys.* V.2.226a23–32, cited in (ii) above, is more

[68] Philoponus makes this point (Philop. *In Phys.* 114.16–32; cf. Simpl. *In Phys.* 186.12–14), though he takes the originating state to belong to the combination of the maternal material and the paternal seed.

[69] Ceasing to be is also problematic for artefacts: one way of destroying (e.g.) a house is by resolving it back into the formless heap of materials from which it came to be; likewise one might melt the bronze of a statue. But a house can also be destroyed by shattering the constituent timbers, bricks, etc. – as a marble statue might also be destroyed. In cases of this type, the state into which the thing ceases to be is not the same as the state from which it came to be. See also point (iv).

[70] ἐν ἀντιφάσει; cf. κατ' ἀντίφασιν at 225a12–14.

upbeat: 'in each of these [change in quality, quantity and place] there is opposition (ἐναντίωσις)' – but even this does not attempt to extend the claim about 'opposition' to substantial change.

VI. Commentary on 188b26–189a10

188b26–30. What is the sense in which Aristotle's predecessors 'suppose things without giving an account'? Presumably it is that they did not engage in the sort of analysis he has given at 188a30–b26.[71] The phrase 'as if compelled by the truth itself' is echoed in *Metaph.* A 3.984b8–11.[72] Despite its somewhat patronising tone, the basic idea is that facts can strike someone without their being able to give a general account of them or of why they are the way they are. 'The things they call principles are the opposites': see the comments on 189a9–10 below.

188b30–189a1. This passage is echoed in the discussion of primary substances as principles in *Metaph.* Λ 1.1069a25–30:

> It was of substance that [our predecessors] sought the principles and elements and causes. Some, our contemporaries, take *universals* to be substances (for the genera are universal, and it is these, rather, that they say are principles and substances – because of their abstract method of inquiry), while earlier thinkers took *particulars* to be substances – for example, fire and earth, but not what is common, body.

The distinction is not exactly the same, however; for in *Phys.* I.5 not only the Platonists (odd and even, the great and the small) but also Empedocles (love and strife) are ranged against the proponents of things more knowable by perception.[73] The inclusion of Empedocles is both surprising and unsurprising. It is unsurprising because in *Phys.* I.5 the distinction rests not on the universal/particular contrast but on that between what is knowable in account and what is knowable by perception: love and strife are theoretical entities posited to account for certain patterns detectable in change.[74] It is surprising because love and strife are neither

[71] See n. 43.

[72] Cf. 984a17–19, where it is τὸ πρᾶγμα, rather than ἡ ἀλήθεια, which is said to have done the compelling.

[73] Other passages in which Aristotle draws related but not obviously identical distinctions include *Phys.* I.1, III.4.203a3–18, *Metaph.* I 2.1053b9–16 (I am grateful to Laura Castelli for drawing my attention to the latter two passages).

[74] Another, related, difference is revealed at 189a1–9, where it appears that 'more universal' is unqualifiedly better than 'less universal' – but where this does not mean that 'more Platonic' is better than 'less Platonic', but rather that the more Aristotelian the better! See the comments ad loc.

the components of things nor the termini of changes: they are principles (in Empedocles' system) in a different way, otherwise altogether ignored in chapter 5 – namely as efficient causes.[75] It could perhaps be that by 'love' and 'strife' Aristotle here means the states of the cosmos during the periods of complete (domination by) love and strife: these states are (plausibly) the termini of cosmic changes. It would be odd to characterise love and strife understood in this way as 'the causes of coming to be', however, unless Aristotle is thinking of these states as the material causes of each other. The phrase 'causes of coming to be' (αἰτίας τῆς γενέσεως) at 188b34–5 might even suggest that (despite the reference to στοιχεῖα at a28) Aristotle is now thinking of *all* these opposites as efficient causes: but in a recapitulation of the earlier account this would be hard to fathom.

'They are different, as indeed they seem to most people to be, yet they are the same by analogy.' The 'as indeed . . . ' remark seems to be a piece of self-congratulation: Aristotle is the one who has detected the key pattern in the apparently chaotic plurality of earlier accounts of principles. One might have expected him to go on to say that they are the same because they all posit opposites as principles; but this would not be a case of sameness by analogy: what he means is made a little clearer in the next passage.

189a1–9. 'For they take them from the same column' must mean, as Ross says, that each of the pairs of opposites they use can be distinguished as positive and negative – though it is a somewhat odd way of saying that. Thus it might be said that they all (in effect) claim that change is between positive and negative termini; but each claim of this sort (invoking positive and negative) is only the same by analogy with the other claims. It is unclear what Aristotle has in mind by 'and worse and better': I suggest he thinks that his 'primary opposition' (between form and privation, though he has not said this) comes at the top of the two columns, so they include all the others, and so that the higher up the columns one manages to get (i.e. the more general one's primary opposites are), the better one's account.

189a9–10. 'That the principles must be opposites is evident.' As I remarked earlier, this – like the claim at 188b27–30 – is a much stronger claim than the one Aristotle started with, and a much stronger one than his arguments (from his predecessors' accounts and ἐπὶ τοῦ λόγου) support; and of course it is stronger than the position on opposites as principles which he is going on to develop in subsequent chapters.

[75] Cf. *Metaph.* A 4.

Physics I.6

Alan Code

I. General Overview of the Chapter

Physics I.5 had argued in support of the view that the principles are opposites. It had also set out opinions of Aristotle's predecessors that attest to this, and offered a classification scheme for the opposites that they had proposed. *Physics* I.6 now furthers his investigation into the principles of changeable, natural things by introducing what he takes to be the next question that needs to be addressed:

> That the principles must be opposites is evident;
> [I.6] what follows next is that we should say whether they are two or three or more. (189a9–12)

Although at this point he takes it as clear that the principles ought to be opposites, the question as to their *number* remains unresolved. For all that has been said in support of the general thesis that the principles are opposites, there could be just about any number. This chapter narrows down the number to just two or three, and does so in four phases:

(1) Arguments for the elimination of the extreme possibilities, (i) that there is only *one* principle and (ii) that the number of principles is *unlimited* (189a12–20).
(2) Presentation of puzzles that result unless, in addition to two opposites, there is a third principle (189a20–34).
(3) Discussion of previous theories that do involve such a third principle (189a34–b16).
(4) Arguments against there being more than one principle that is a substratum, or more than two principles that are opposites (189b16–27).

Given the chapter's initial assumption that it is clear that the principles are opposites, Aristotle straightaway argues in (I) that there are *at least* two principles. This leaves it open whether the complete set of principles should include more than just two opposites, and by the end of the chapter he will also have presented his main considerations in favor of there being *three* principles. However, before turning to arguments for a third principle, and subsequently to arguments for no more than three, he first sets out arguments against there being an *unlimited* number. Next, having eliminated the view that they are unlimited, phase 2 is concerned to show how the main conclusion of the previous chapter leads to difficulties that require the introduction of a third principle. In particular, there is reason to hold that the *substratum*, what underlies the opposites, is a principle that is not itself an opposite.

His arguments for taking the substratum to be a principle are crucial to the way in which this chapter deals with its main question as to whether the number is two, or three, or unlimited. If every principle were an opposite, then the number of principles would simply be the number of these opposites; and if there could not be four or more opposites that enjoy that status of a principle (as argued in (4)), there would in that case be *exactly two* principles. However, if the substratum for those opposites is itself a principle, and yet not itself an opposite, there would have to be *three* principles (assuming no more than these *two* opposites, and no more than *one* substratum for them).

In what follows each of the four main phases will be discussed. However, in order to highlight both their relations to each other and their collective role in the larger context of the progressive clarification of the principles, much of the discussion of the later phases will be "front loaded." That said, positing the substratum as a third principle distinct from a pair of opposites is hard to reconcile with the chapter's initial starting point, his claim that it is clear that the principles must be opposites. The previous arguments in I.5 are treated as though they should be preserved, and in the course of his presentation of arguments for the conclusion that the *substratum* is a principle he does not adopt the view that this new conclusion should be rejected in order to save the earlier conclusion. In fact, as we shall see, he says that the positing of this third principle is necessary if one is to hold both the earlier view of I.5 and the new account he provides in phase 2. His take on this is perplexing since the substratum for a pair of opposites is itself opposite to neither of them, but rather is "*another nature* for the opposites" (189b28–9; italics added for emphasis), and it is hard to see how that is compatible with the claim that the principles are opposites, and even harder to see how it is required in order to preserve that view.

Accordingly, unless there is one way in which this third item is distinct from the opposites, but also some *other* way in which, for the purpose of counting items in the ontology, the substratum is not, his new conclusion flatly contradicts the conclusion drawn from the earlier arguments. The acceptance of both views will require a nuanced understanding of the conclusions of his arguments in I.5 and I.6. He will need to draw relevant distinctions in terms which would enable him to show how to understand them both in such a way that both are in fact correct, at least in suitably modified versions that nonetheless manage to retain the central insights of each.

Accordingly, the present chapter brings the inquiry to a juncture at which further progress requires resolution of an *aporia*. A grasp of the distinctions needed for this resolution is what enables one to give a precise account both of what the principles are and of how many they are. This is why, by the completion of the four main phases of I.6, the inquiry has led to a point of extreme puzzlement:

> Well, then, that there is neither one element, nor more than two or three is clear; but whether there are two or three, just as we have said, is very difficult. (189b27–9)

As he transitions to his own positive account in I.7 the number of principles that are elements has been narrowed down to either two or three. The options still in play are (i) there are just two principles, both of which are opposites, or (ii) there are three principles, two opposites and one substratum (that is not an opposite). On either of these options the results of I.5 have secured that at least two opposites are principles, and since phase 4 showed that there are at most two opposites, and one substratum, it has also shown that there are at most three principles overall. Let us now consider in more detail how he arrives at this point.

II. Initial Elimination of Extreme Possibilities

Although Anaxagoras has nothing to contribute to the introduction of the third principle invoked in I.6, the main aims of the chapter will not be completed satisfactorily without considering him once again. Even if the arguments in I.4 have ruled out the view that the set of principles includes an infinite number of uniform stuffs, the general thesis that the principles are opposites still leaves open the possibility of an unlimited number. The first phase of I.6 does not show that the number of principles is small, much less as small as two or three. The main aim of this early phase of I.6

is to give Aristotle's arguments against extreme options before turning to the arguments in support of a principle that is not an opposite, and next to the reduction of all opposites to a single primary opposition. Although a couple of the arguments in this early part of I.6 are formulated in such a way as to bear upon the number of principles that are *opposites*, the other arguments argue against an infinite number of principles of any sort, whether or not they are opposites.

Phase 1 initiates its sequence of arguments designed to eliminate the extreme possibilities (i) that there is just one principle, and (ii) that the principles are unlimited in number, as follows:

> Well, they cannot be one,[1] because the opposites are not one, and they cannot be unlimited, both because that which is would not be knowable, and . . . (189a12–13)

As is more generally the case with the arguments in this chapter, this conclusion will require qualification once the distinctions needed to solve the *aporia* are in place. After briskly dispensing with (i) in 189a12, he goes on to initiate a series of considerations against (ii). The first argument against an unlimited number of principles is his compressed statement that being, or "that which is," would be unknowable if there were infinitely many principles. This passage does not provide any further premises concerning knowability that would help one see why an infinity of principles would rule out knowability. However, I.4 has already used the claim that the infinite as such is unknowable when arguing against Anaxagoras,[2] and the earlier knowability arguments generalize to other theories as well. The present passage could rely on the claim there that scientific knowledge of X requires knowledge of each of the elemental constituents of which X is composed, as well as knowledge of how many they are.[3] In that case, something constituted out of an infinite number of ingredients would itself be an object of scientific knowledge only if the knower had some type of knowledge of each of an unlimited plurality of ingredients.

To rule out the knowability of something with infinitely many principles one would need to assume additionally that it is impossible to have such knowledge of each member of an indefinite plurality. A view such as Anaxagoras' (as opposed to that of Empedocles) would thereby preclude

[1] With γάρ (here untranslated) he initiates the discussion of the questions just said to be next.
[2] 187b7ff.
[3] 187b11–13. One variant of this kind of interpretation is Ross's claim that this argument depends upon the still earlier passage 184a12–14.

the knowability of natural things because it treats them as composed of an unlimited number of constituents. Such an argument would not, however, rule out *all* views that countenanced an infinite number of principles. A theorist who, unlike Anaxagoras, holds that any given object is constituted out of just a finite number of principles could nonetheless maintain that the total number of principles is unlimited. On one such view, for instance, each kind of knowable thing is composed of just a limited number of principles, but there is no limit to the number of kinds that are exemplified in the world.

In any case, the laconic knowability argument in I.6 is perfectly general, and does not point to features that are specifically connected with the thesis that *opposites* are principles. However, the second argument against an unlimited plurality of principles utilizes the premises that (i) in the case of any single genus (γένος) there is a single opposition (ἐναντίωσις) and that (ii) substance (οὐσία) is a certain single genus:

> There is a single opposition in the case of any single genus,[4] and substance is some one genus. (189a13–14)

This would militate against there being an infinite number of oppositions for the kind substance. If one were to add to this the claim that natural things are substances, given that substance is some one genus of being, this would mean that for the most general kind to which natural substances belong there is just a single opposition. Although there could not be an unlimited number of oppositions that characterize their most general kind at this maximum level of generality, this is not to say that there is one and only one opposition for each of the sub-kinds of the genus, or that there are only two opposites possessed by physical objects.

Later in I.6 he will argue at 189b22–7 that there cannot be more than one *primary* opposition.[5] In retrospect this helps further clarify the earlier argument. Despite the fact that there are numerous oppositions, there is just one opposition that is first, or primary, and this is the single opposition for the kind substance. The two opposites that have the status of *principles* of substance are the members of this single *primary* opposition. This later passage gives a somewhat clearer account of what it is for there to be a single opposition for a single kind, and considerably more clarity will

[4] There is an opposition for the genus substance, but neither that opposition nor the opposites that it contains are themselves *in* that genus, included among its members. Hence it is best to construe ἐν + dative here as "in the case of . . ."; cf. Plato's *Laches* 190d4–5: ἡ ἐν τοῖς ὅπλοις μάθησις.

[5] The primary opposition for a genus is said to be a positive state and its privation at *Metaph.* I 4.1055a33.

be provided still later when the opposites are identified as *form* and *privation* in the analysis of I.7.

For now, though, let us consider what we are in a position to say about the oppositions for the kind substance without looking ahead to further developments. Take the example of something coming to be a substance of a specific kind, such as a horse. In this case, for what comes to be, its *being what it is* is *its being a horse*. What here plays the role that will eventually be assigned to form is this positive condition: *its being a horse*. Furthermore, there is something which, not being a horse, comes to be a horse. What here plays the role that will eventually be assigned to privation is the privative condition: *its not being a horse*. This is not simply *not being a horse*. There are all sorts of things that are not horses, and all sorts of cases of not being a horse, that would not be appropriate starting points for the generation of a horse. What becomes a horse is, prior to the generation of the horse, in the condition of something which is not yet a horse, but by being affected in just the right way, changes in such a way as to be a horse. The privative condition that is a starting point from which a horse comes to be, the relevant way of *not being a horse*, is precisely the *not being a horse* of that which becomes a horse.

There is a way to subsume these specific opposites under a maximally general pairing of being a substance and not being that. Speaking at the level of the genus substance, the principles of a (generable) substance S are being and not being S. Returning to the previous example, there is a single item in which the role of *not being a horse* and the role of *that which becomes a horse* coalesce. Likewise, if we consider the situation in which a horse has now come to be, the positive condition of *being a horse* is a condition of that which, although formerly not yet a horse, has now become a horse. To speak again in general terms, it is the *being what it is* of what has become precisely *that*.

If we adopt the point of view of I.5, and consider the opposites in connection with substance, we are already in a position to see that the starting point for the generation of a substance S combines the roles of becoming S and not being S. When phase 2 explicitly brings in the substratum as what becomes something, Aristotle's investigation has come one step closer to disentangling the two roles played by a single item, the item that becomes something new. To make this progress the general thesis that there is a single opposition in the case of a single genus needed to be considered in relation to just one kind, the single kind substance.

In the present chapter Aristotle's inquiry itself fails to distinguish (i) cases in which there is something that becomes X *without* being also that out of which X comes to be, from (ii) cases in which what comes to be something X is without qualification that out of which X comes to be. This first pattern, (i), applies to every case in which an individual substance undergoes an accidental change. In accidental change what comes to be does so out of what is opposite to it, but it does not come to be out of the individual substance that loses one opposite and acquires the other in the process. By way of contrast, the second pattern, (ii), applies to the generation of a new substance, but not to accidental change.

Natural objects change in various ways, and prior to getting to the distinctive differences between these various ways of changing, the focus of the investigation in I.6 is on the principles for the kind *substance*, but without specifically distinguishing substantial change from other types of change. The opposites for the kind substance will turn out, in phase 4, to be the principles for anything that comes to be, and the principles of substance are thereby the principles in the most general, or comprehensive, manner possible. Speaking at a relatively low level of generality one may identify the principles involved in the generation of a horse, or a dog, but whatever is a horse or a dog is also an animal, and also a living thing, and also a substance. The very same items that are in the kind *horse* are also correctly classified using kinds of increasing generality, and ultimately may be classified simply as a substance.

The next two arguments do not make use of the claim that opposites are principles. Unlike the preceding argument, which associates just a single pair of opposites with a single kind, these considerations purport to do no more than show that the number cannot be unlimited:

> and because it is possible [to explain] on the basis of a limited number, and it is better [to explain] on the basis of a limited number – for instance as Empedocles does – rather than on the basis of an unlimited number; for he thinks that he explains just as much as Anaxagoras does on the basis of an unlimited number. (189a14–17)

This compressed argument does not argue for the claim that it is possible to proceed with a limited number. This is immediately linked with the further claim that it is also *better*, and perhaps the two features may in fact be interconnected. He reminds us that the view of Empedocles is better than that of Anaxagoras (stated earlier at 188a17–18), and at least part of the reason why it is better is that Anaxagoras' view of an infinite number of uniform stuffs is not even possible. The implicit thought in the present passage might be that it is impossible to explain in terms of an infinite number of principles, but since it is not impossible to explain in terms of a finite number, it is also better. This reasoning favors a limited number.

For this kind of argument to be useful there should be at least some support for the claim that it is impossible to proceed on the basis of an infinite number. Merely arguing against an infinite number of uniform *stuffs* does not by itself also show that a theory with an infinite number of opposites is worse than one with a finite number. The earlier argument at 189a11–12, which alleged that an infinite number of principles would render *that which is* (i.e. *being*) unknowable, could be used to help secure the desired result. In that case the next two arguments could be seen as drawing out further consequences of the earlier argument. Theories with an infinite number of principles are impossible, because were there an infinite number of principles this would preclude the knowability of the beings that fall within the scope of a science. This would explain at least one aspect of the goodness of finite theories. They are good, or better, precisely because when a theory is finite it is at least not impossible for that theory to yield knowledge.

The final consideration adduced in support of the finitude of principles is that some opposites are prior to other opposites:

> Furthermore, some opposites are prior to others, and some come to be from others; for instance, sweet and bitter and pale and dark, but the principles must always remain. (189a17–20)

There is a textual question as to whether we should follow Ross and read ἐξ ἀλλήλων (from one another) with FI at 189a18 instead of ἐξ ἄλλων (from others) with EVS. As 188a27–8 already has indicated, principles are neither from one another nor from other things. Accordingly, primary opposites are neither from others (ἐξ ἄλλων) due to their primacy, nor from one another (ἀλλήλων) due to their being opposites. The pairs given as examples here (sweet/bitter and pale/dark) are not plausible candidates in which one *pair* arises out of the other; and although pale comes to be out of dark, and vice versa, given the quite general thesis that opposites come to be from opposites, reading the text in this way loses the contrast between these cases and the primary cases of a17–18. These considerations lend support to reading ἐξ ἄλλων here, and accordingly an opposition such as pale/dark is derivative and non-primary, and there is some other opposition prior to it, out of which it comes to be. On this reading, the claim that some opposites are prior to others leads naturally to the subsequent claim that different oppositions (different from those that are primary) are generated out of other things (out of primary things).

The ones generated out of something else are not primary, and they are not primary simply because they are generated out of something else.

Read in this way the examples bitter/sweet and pale/dark arise from causally more fundamental oppositions, and hence are "from others." It is not explicitly said how such considerations support the idea of there being a *limited* number of principles. Suppose that some opposites are prior to others in this way. Could there nonetheless still be an unlimited number of primary, primitive principles? To address this question one might consider taking these two oppositions as candidate pairs of opposites that are prior to other opposites. Since he takes colors to be ratios of the opposites pale and dark, it is not difficult to construe pale and dark as prior to ratios of pale and dark. That is, instead of taking bitter/sweet and pale/dark as examples of opposites that are themselves generated out of other opposites, on this reading they are treated as opposites that are prior to other things (i.e. other colors or flavors). Hence there would be different opposites (i.e. different from such allegedly primary opposites) from the range of colors or flavors that are generated from other opposites (such as pale, dark, bitter, and sweet). However, this reading implausibly treats intermediate properties as "opposites," and may also fall foul of the principle that for each kind there is a single opposite. Although this would explain how they are prior to intermediates, it does not yield a sense in which they are prior to *other opposites*. At best it engages an opponent who mistakenly classifies intermediates in this way, and so it would be of little use in developing Aristotle's own views.

It is preferable to interpret the passage as making the perfectly general and (for Aristotle) correct contrast between derivative opposites such as pale and dark which, although prior to the intermediates in the genus, are not the primary opposites for the kind color. They are members of the kind color, as are the intermediates, but they are defined in terms of a prior pair of opposites that are not members of that kind. The opposites "penetrating" and "compressing" are the differentiae in the definitions of pale (a penetrating color) and dark (a compressing color), and the genus is of course *color*.[6] Accordingly, there are *two* oppositions for the kind color, only one of which is the primary opposition. The single opposition for the kind color is a pair of opposites that are not themselves members of the kind, and are prior to the derivative opposites (pale and dark) that are in the kind.

[6] διακριτικὸν χρῶμα, συγκριτικὸν χρῶμα; see *Metaph.* I 7.1057b2–34.

In this argument the priority in question is explicated in terms of certain opposites being derived *from* (ἐκ) others, and not explicitly in terms of levels of generality. Although this chapter is concerned to find the greatest level of generality at which the principles may be described, in doing this it considers how many specific principles are required for the individual cases at the lower level of generality from which it generalizes. If the eventual conclusion, in the subsequent chapter, is going to be that a trio of form, privation, and substratum are the elements, then for any individual change there is an entity that plays the role of a form, an entity that plays the role of a privation, and an entity that plays the role of a substratum.

If the inquiry starts with examples of opposites in lower kinds such as color as a kind of quality, or animal as a kind of substance, that are less general and nearer to sense perception, one can then proceed to search for the opposites that are principles for *everything* whatsoever that comes to be by nature. Suppose, for instance, that bitterness and sweetness are the *primary flavors*, and pale and dark the *primary colors*, all other flavors and colors being ratios of these primary properties. These are the primary opposites that are present in the kinds flavor and color, respectively. Even if there are indefinitely many ratios of pale to dark, and so indefinitely many colors, there is nonetheless just a single opposition within the kind color. However, although this is primary in relation to the other members of the relevant genus, in each case there will be some other prior opposition that falls outside the kind, but figures in the determination of the primary qualities for that kind.

The number of derivative principles within the kind color is limited (whether or not the number of colors is limited) in that for each one kind there is just one primary opposition, as required by the proposition at 189a13–14, which is prior to the opposition within the kind and itself outside the kind. The opposites within the genus or kind are in each case derivative, and generated out of just two opposites. (That the intermediates will be generated out of the primary opposites that are in the kind is a separate point.) This does not establish the finitude of principles that are opposites without the further assumption that there are a finite number of *summa genera*, or that the principles that are opposites are the opposites for the single kind that is primary and prior to the other *summa genera*. He has already claimed that substance is a single kind in an earlier argument in I.6 for a limited number of principles, and associated with it a single opposition. The present argument complements that earlier argument by making it explicit that some of the oppositions are derivative, and prior to them are the oppositions from which they are derived. If this is interpreted along the lines I have been suggesting this helps explain why there is just a single opposition for a kind that is prior to everything within that kind, and why such a prior opposition is itself outside the kind in question.

He concludes this phase with the additional premise that in every case it is necessary that the principles *remain* (a19–20). Later, in I.7, one of the elements is a substratum that persists (ὑπομένει, 190a17ff.) through change, but this does not seem to be what he has in mind here. Our present passage is not concerned with a substratum that remains, and in fact occurs before he even brings in the arguments in support of a third principle, a substratum for the opposites. At this earlier stage he is still operating under the assumption that the principles are opposites, claiming that some opposites are prior to others. Only those *opposites* that remain are principles, and this does not invoke the persisting substratum for change between opposites.

Rather, the claim is offered to support the combined thesis that (i) some opposites are prior to others, and (ii) some arise from others. Given the interpretation of priority just sketched above, the second claim (ii) turns out to support the first claim (i) in the following way. It is because pale and dark, for instance, arise from the penetrating and the compressing that the latter opposites are prior to the former. In this example the principles that "remain" are the differentiae of the two primary colors. They are contained in the primary colors as elements mentioned in their definitions, and because all of the other colors are ratios of pale and dark, they also are present in the intermediates. All of these, pale, dark, and intermediates, are derivative in that they are "from" or "out of" things that are prior to them. Their prior principles are what remain when they are analyzed into those things out of which they are generated.

III. The Need for a Non-Opposite Principle: Aristotle's Predecessors and This Underlying Principle

As already noted, the chapter's concluding puzzle about whether the principles are two or three arises from the combination of (i) arguments in support of positing the substratum as a principle with (2) the earlier arguments that purport to show that the principles are opposites. We have already discussed in general terms the difficulty of combining them both in a single view. Let us now consider how the chapter's arguments support this third principle, that which underlies.

The first of these arguments, unlike the *logical* considerations in phase 1, makes use of a specific *physical* requirement for change:

> There is some reason for not making them only two; for one might raise a difficulty about how either density is of such a nature as to make rarity something, or rarity make density something; likewise too for any other opposition, for Love does not gather Strife and make something from it, nor does Strife make something from it, but both <make> some other third thing <into something>. And some assume even more things from which they construct the nature of the things that are. (189a21–7)

The first case, that of density and rarity, recalls the treatment in I.4 of natural philosophers who posit a single material body as substratum. Many other things (other than the single material substratum) are generated by density and rarity. The second case, that of Love and Strife, recalls Aristotle's subsequent treatment of Empedocles, a representative of pluralist views, and the positing of many material bodies as principles. However, this is not a puzzle specifically for materialists. It is addressed to anybody who accepts that what comes to be does so from what is opposite to it, and posits (at least) *two* principles that are opposites. Despite numerous disagreements among such theorists as to which pair of opposites is fundamental, or as to which body or bodies ought to be posited, it would be very puzzling if any of these theorists were to maintain that the only principles were a favored pair of opposites.

If there were not some principle that gets made into something by the opposites there would not even be naturally generated items, and so there would be nothing for these two opposites to be principles of. But if there is even one generable object, its existence is due to the fact that something was made into it. In that case, the generated item would have as one of its principles that item, whatever that is, that was already there in the first place, and is precisely that which has been made into this generated item.

This argument indicates the reason why the complete set of principles could not consist of just a single pair of *opposites*: one member of an opposition never makes the other member into anything at all. Any theory that confines the principles to just two opposites has no room for an additional principle that gets made into something, but if something is to be made, an additional principle is just what the natural scientist needs if neither of the two opposites can play the role of that which is transformed into something. This does not show that the opposites are not principles, but it does provide a reason to hold that they could not be principles on their own, independently of any other principle, or principles.[7] However, in that case it does show that not all principles are opposites, and that would seem to support the view that there is at least one principle additional to, over and above, whatever opposites are principles. Accordingly, on the assumption that there are at least two opposites that are principles, these opposites are principles together with the third principle, the item that is made into something. So, the thesis that the principles are opposites needs to be conjoined with the thesis that for generation to occur at all there must be something that is affected and made into the new item,[8] and that which is affected is a principle that is not one of the opposites.

Aristotle does not think that his arguments in favor of positing a substratum as a principle refute or replace the main result of I.5, the thesis that the principles are opposites.[9] In fact, we are explicitly told that somebody who wishes to keep *both arguments* will need to posit this third item:

[7] This view is clearly stated by Simplicius, *In Phys.* 199.8–10.

[8] When Aristotle speaks of opposites that make (ποιεῖν) an X out of Y, or of some Y that is affected (πάσχειν) by opposites in such a way that an X is produced, or uses related vocabulary, this might be a case of opposites serving as the efficient cause of the production of X, or it might instead be a case in which the presence of a property is that by virtue of which some subject Y is X. For instance, in *GC* I.1 we are told that two Empedoclean elements are efficient causes. These elements are Love and Strife, and are not the corporeal elements that are the basic ingredients out of which generated things are constituted. However, as phase 4 makes clear, I.6 is concerned with the number of *elements*. If so this chapter would not be concerned with a principle that is not an element, or whether there are principles in addition to the elements. The topic the chapter considers is whether the elements are a pair of opposites, or a pair of opposites and a substratum. When he says in I.7 that it is impossible for opposites to be affected by each other (ὑπ' ἀλλήλων γὰρ πάσχειν τἀναντία ἀδύνατον, 190b33) this too need not yet be a claim about efficient causality. An instance of this claim would be that the hot and the cold cannot be affected by each other. This could be construed not as a claim to the effect that the hot cannot act as an efficient cause on the cold in such a way as to produce something hot. Rather, when something is hot, it is hot by virtue of the hot being present to it, but it is impossible that it is the cold that is hot by virtue of the presence of the hot to the cold. This would be because the subject that has the property of being hot is somehow something distinct from both the hot and the cold. Its way of being distinct is to differ in being, but somebody who has not yet solved the *aporia* from I.6 would not be able to explain the difference between the property of heat on the one hand, and both (i) the hot thing that acts as an efficient cause on the subject that is (prior to a change) cold and (ii) the hot that is produced by the change. Aristotle's discussion prior to I.7 of his predecessor's views of opposites as principles lumps together efficient causes that are opposites with opposites as elements or ingredients of what is produced.

[9] For a different interpretation of the relationship between I.5 and I.6 see the excellent article by Kelsey (2008). He argues that the considerations in support of the third principle "systematically undermine the reasons given earlier in I.5 for thinking that contraries are principles in the first place" (186). For reasons given in the main text I think that the new arguments are not privileged over the I.5 arguments, and that the chapter ends in genuine perplexity as to how to keep the former and latter together. Kelsey insightfully connects "opposites come from opposites" with a destructive effect on what is there in the beginning, and associates the argument for the substratum as a third principle with a constructive effect. I fully agree that in I.7 these "seemingly competing conceptions" are reconciled and incorporated into a single, unified theory.

> Consequently, should one hold that both the earlier account and this account are true, it is necessary, if one intends to preserve them both, to posit some third thing. (189a34–b1)

He does not yet indicate, as he will later in the chapter, that the introduction of the third principle will cause problems for his earlier arguments. To the contrary, the truth of the earlier thesis is what required the introduction of the substratum as a principle in the first place. It is important to see that the difficulty with which he starts phase 2 has been resolved by the introduction of the substratum. This *aporia* plays a distinctly different role from the *aporia* at the end of the chapter. So, part of the initial motivation for the third principle is that it saves the thesis that the opposites are principles. That said, once the third principle has been introduced it needs to be incorporated into the account of the principles that has progressively emerged, and this cannot be accomplished simply by tacking it on to the earlier results. Modifications, refinements, and adjustments will be needed for both sets of arguments. The concluding *aporia* about whether the principles are two or three is both a reflection of how hard this is and the challenge that serves as the transition point to his own positive account of the principles. Nonetheless, before moving directly to I.7 Aristotle does not pause to explain why one ends up with a troublesome *aporia* about the number of principles. That said, once it is viewed from the perspective of the emerging account of principles that he has been extracting from the commitments of those engaged in the study of nature, no explanation is necessary. The tensions within their own project are highlighted as their commitments are extracted from their views, but the problem has been there all along. It has now been brought out, and if he is to move forward with an internally coherent enterprise it now needs to be resolved. Through the route his inquiry has taken up to this point this is the first time this foundational problem for the entire project of natural philosophy has been properly formulated.

The remainder of phase 2 goes back to discussing the principles in terms of general logical considerations about predication, categorial being, and substance that are brought to bear on the primacy of substance. Although the introduction of the third principle solves phase 2's initial *aporia*, thereby giving a, or the, rationale for the natural philosopher for this posit, there are also logical considerations that support taking what underlies opposite properties as a principle. In fact, these logical considerations do not introduce the underlying substratum for a generable substance, but rather are about the logical subject for accidents. As Aristotle tells us in *Metaph.* Z 13.1038b4–6, that which underlies does so in two ways. On the one hand, there is that which underlies in the way an animal underlies its affections. This is the way in which an individual substance is a *subject* for accidental predication. On the other hand, there is that which underlies in the way matter underlies fulfillment, or *entelecheia*. This is never a particular, individual substance (except potentially), and is not the subject for accidental predication. It is instead the substratum for substantial change and the generation of a substance. Accordingly, these remaining arguments are not even about the introduction of the substratum as a principle. Although I.7 will draw distinctions that enable one to see this, Aristotle's predecessors conflated the two ways of underlying, and at this stage in I.6 he discusses the third principle in a way that confuses the two ways of underlying.

Phase 2 brings in these logical puzzles as additional points that the natural philosopher needs to take into consideration in light of the introduction of the third principle in this way:

> In addition to these points, one might raise a further difficulty over this, should one not posit another nature for the opposites . . . (189a27–9)

What the immediately following words, at a29–32, are about to add to the preceding arguments is that the required third thing is both a *substance* and is a *principle* that is prior to the opposites that characterize this substance by being predicable of it:

> We do not see the opposites to be the substance of any of the things that are, and the principle ought not be said of some underlying thing. For <if so> there will be a principle of a principle, for what underlies is a principle, and is thought to be prior to what is predicated. (189a29–32)

The third item posited in the previous phase, something that is affected by one of the opposites, is here viewed not from the perspective of this role in natural change, but mistakenly from the more general perspective as the subject of accidental predication, where viewing it as a subject of this sort is to view it as a substance.[10] Speaking quite generally, the opposites

[10] Here I translate ὑποκείμενον as "subject," rather than either "underlying thing" or "substratum," in order to mark the difference between the role of a subject of predication in logic and the role in natural science of a substratum that is affected and becomes something. Although he does not use this Greek term in the immediately preceding arguments in a22–7, they introduce the third item by invoking the need in natural science for something that underlies change. By contrast, the additional puzzles (πρὸς δὲ τούτοις ἔτι) concern the need in logic for something that underlies the predication of opposites, where the former is a substance that serves as the subject of which these non-substantial properties are predicable. Whereas the matter that becomes something is a ὑποκείμενον in the first of these ways, that of which non-substantial properties are predicable is a subject in this second way, and any substance that comes to be out of matter is here treated as an

that are predicated of a substance are not what that substance is. The substance is something the being of which is different from the accidents that characterize it, and so neither of them is the substance of that substance which serves as their subject.

This argument is not put in quite these terms because he has not yet said, as he is about to, that the opposites are not substances. Here he simply says that "we do not see the opposites to be the substance of any of the things that are," and this is certainly true if we are considering only those opposites that are predicable of individual substances. In the absence of a distinction between the two ways of underlying one might take the fact that none of the opposites that are countenanced in the *Categories* are the substance of their subjects to settle the issue.

Of course, if one of them were the substance of the subject, it would be one of its principles. But in the present case the reverse is true, and the subject is the principle for what is predicable of it. The subject of which these opposites are predicable is a principle for them because quite generally, and not just in connection with change and the philosophy of nature, accidental properties are dependent upon the substances that are their subjects. Given the way (or perhaps ways) in which non-substantial properties depend upon subjects (whatever that way is), they too need a third item that underlies them, a substance. When the subject is a particular substance, the opposites simply cannot be substances. But since they are predicable of a substance, they still enter the ontology and have the status of beings. In this way the third item is a principle for them, and as such is prior to the opposites predicable of it. If the opposites that are predicated of substances are principles, they cannot be principles of substance, and they certainly are not principles of the substance of which they are predicable. Although he is not in a position to say this yet, they are principles of accidental composites.

Thus there is now a second reason in favor of the claim that opposites require the existence of a principle that is not an opposite, namely what underlies them. But, since this invokes a different way of underlying, the

example of a logical subject for non-substantial properties. That said, the requirement that the substratum that is affected is something different from the opposites does not resolve the issue as to whether this third item is different in such a way that there are three *principles*. Having argued that the principles are a pair of opposites and a substratum, one still needs to articulate the way in which for a natural change what is there at the start is that one thing which, by being affected in the appropriate way, is what becomes something other than what was there in the beginning. However, at this present aporetic stage the manner in which what underlies is entangled with the opposites is yet to be sorted out, and the relevant distinction between the two ways of being a ὑποκείμενον has not yet been properly distinguished.

natural scientist will need to consider how this new argument relates to the first argument for positing the substratum as a principle. In order for *these* opposites to be principles at all, given the dependent nature of properties, they require a subject as a prior principle. However, nothing like this is true when what underlies does so as matter. In those cases, it is the form that is prior to what underlies, and not the other way around.

His second argument could be clarified by showing that this third item belongs to a different category from that of the opposites, and he is about to do just that. The subject belongs to the category of substance because it is the subject of predication. The next claim that he makes is that a substance cannot be opposite to a substance. This allows him to infer that the opposites are not substances, and with that premise secured he concludes this phase with two brief arguments that are in effect simply further elaborations and clarifications of the earlier argument that the subject is a principle prior to the opposites predicable of it.

The final arguments in phase 2 make use of the claim that the opposites that are predicated of substances are not themselves substances to raise two unanswerable questions about them:

> Further, we say that a substance is not opposite to a substance. Well then, how could a substance be <constituted> from non-substances?; or how could what is not a substance be prior to a substance? (189a32–4)

So, given that no substance is opposite to a substance, it follows that opposites are not substances. Among other things, neither of the opposites predicable of some substance X are the substance of X. This leads to two questions, one about the composition of substance, the other about the priority of the constituents. First, given the assumption that *what something is* is constituted out of its principles, whatever is from opposites would have opposites among its constituents. Hence, were a substance to have such a constituent, something that is not a substance would be a principle of substance. The second point shows why this could not be the case. Since a principle is prior to that of which it is the principle, if a non-substance were a principle of a substance it would be prior to that substance. However, it is not possible for what depends upon a substance to be prior to it, and so what is not a substance could not be prior to a substance.

Having just given his own arguments to show that there must be some underlying nature for the opposites, we are now told that this is just what some predecessors in fact asserted:

> ... just as those saying that the universe is some single nature affirm; <saying that it is>, for example, water or fire or what is intermediate

> between them; but it is thought that what is intermediate is more so, for fire and earth and air and water are already entangled with oppositions. (189b2–5)

These are the monists that had been introduced right at the start of I.4. Notice also that although, not being a material monist himself, he is in no position to adopt the view that this single nature just is that which is intermediate between water and fire, we are told that this candidate is better than any of the four (fire, earth, air, and water) because these four are "already entangled with oppositions." This is a clue that on Aristotle's own view the account of what this third principle is should *disentangle* it from the opposites that characterize it. The preceding phase had shown a basis for holding that opposites are not the substance of the underlying nature, and that the underlying thing could not be constituted out of opposites. The opposites are not what the underlying principle is, and, whether construed as the substratum for substantial change or the subject for accidental change, it should not be "entangled" with them.

IV. Argument against More than Three Principles

When Aristotle reaches the final phase he has not only reduced the number of principles to a finite number, but thinks he has reason to limit the number to either two or three:

> Well, there would seem to be some reason to assert that the elements are three for those who are looking closely on the basis of these and other such considerations, just as we said, but that they are more than three no longer <would there be>. (189b16–18)

He takes it that earlier parts of I.6 have given reasons in favor of a third principle, and also for there being at least two opposites among the principles. This phase is intended to show that there cannot be more than one substratum for one opposition, and there can be only one primary opposition, and so only two principles that are opposites. He starts by arguing against an additional substratum:

> For the one [underlying thing] is sufficient to be acted upon ... (189b18–19)

The substratum is introduced in an earlier section of I.6 as whatever it is that the opposites affect, and because this cannot be one of the opposites there is a need to posit some principle other than the opposites for them to affect. However, once a single substratum has been posited, there is no longer any need to posit anything further to play that role. One substratum is enough. Thus the number of principles need not increase by the addition of a second substratum. If there is just one substratum for a single opposition, the only way there could be more than one principle that is a substratum is for there to be two pairs of opposites that are principles. Accordingly, he argues next against the addition of an additional opposition:

> ... and if being four <opposites> there will be two oppositions, some other intermediate nature will need to belong separately to each; but if they are able to generate from one another, being two <oppositions>, one of the oppositions would be superfluous. (189b19–22)

This passage starts by showing how a second substratum would have to be motivated. If there were two more opposites (and hence four opposites and two oppositions) then there would be two distinct "intermediate natures," one for each opposition. The intermediate nature for an opposition is the substratum that underlies both members of the pair, and the claim he is examining is that there would be two distinct substrata on the condition that there are two oppositions.

If indeed a single nature is sufficient, this would already rule out the positing of two such natures, one for each opposition. After all, generation takes place when something that is already there is acted upon in such a way that it is made into something. This is precisely the role of the substratum. Hence, were there two distinct substrata, then two different items would have come to be, and there would not be a single change that involved two substrata and two oppositions. In that case, two different substances would have been generated, and no one substance would have more than three principles. Speaking generally, the principles would be two opposites and one substratum, and so there would still be three principles (and not six).

The next lines, which consider and reject a possible move, are difficult, and a number of interpretations have been suggested. The structure of the passage suggests that this move attempts to show how there could be four principles that are opposites for a case in which there is just one substratum. The new proposal is that there could be two oppositions for a single substratum if these oppositions can generate from each other (ἐξ ἀλλήλων δύνανται γεννᾶν). These two oppositions could somehow operate on a single substratum.

There is controversy as to what this way might be. Philoponus rejects two possible readings of ἐξ ἀλλήλων, and proposes that Aristotle means that each of the oppositions is able to generate everything independently.[11] He rightly claims that it does not mean that one of them generates some things, and the other generates certain other things, or that the mere combination of the oppositions generate everything. However, to get his preferred reading, for the two oppositions to generate ἐξ ἀλλήλων is for each to generate from its own two opposites. Against this, the two oppositions are the subject of δύνανται γεννᾶν, and so the oppositions themselves (not their component opposites) are the "each other" (ἀλλήλων). In some way or another, the oppositions mutually generate something by affecting the substratum in such a way as to make it that thing.

What one needs to find, then, is a way of understanding how they could *mutually* generate something. He does not say that they mutually generate *each other*, but rather that they mutually generate (something) from each other. Fortunately, we already have considered a way in which two distinct oppositions for a kind generate something, but not independently. The pair pale/dark and the pair penetrating/compressing jointly generate an intermediate color in a way that involves both intrinsically. First, the opposites pale/dark are generated from the prior opposites penetrating/compressing. One way in which the latter opposites are prior to the former is that penetrating is included in the account of pale, and compressing in the account of dark. These prior opposites are not colors, and so are not the primary colors. The opposites pale and dark, although posterior to the opposites that serve as their *differentiae*, are the primary colors. Next, intermediate colors are generated from the pale and the dark. The latter are included in the accounts of the former, since the intermediates are defined as ratios of the primary colors. However, since the primary opposites penetrating/compressing are part of the being of the primary colors pale/dark, these primary opposites are also part of the being of the intermediates – they "remain" in the generated products, the intermediate colors. The *differentiae* for colors are either the primary opposites or ratios of primary colors, and the elements of a color are its genus and the *differentia* that makes it the color that it is. The genus is a subject (although of course not a countable particular), and these two oppositions are able to generate jointly an intermediate color by the way they modify this subject.[12]

[11] Philoponus, *In Phys.* 149.23: εἰ καὶ ἑκατέρα αὐτῶν δύναται καθ' ἑαυτὴν πάντα γεννῆσαι.

[12] Simplicius criticizes a similar reading of the phrase ἐξ ἀλλήλων given by Porphyry (*In Phys.* 206.3–8). According to Porphyry, Aristotle is considering theories according to which one of the two oppositions generates the other. His example, directed against the theory that hot/cold and rare/dense are both primary oppositions, is that the hot is generated from the rare and the cold from the dense, and so the pair rare/dense will be sufficient for generation. Consequently, the other

Although the pale and the dark are principles of the intermediates, the penetrating and the compressing are prior principles both of the primary colors and of the intermediate colors. Consequently, since the principles for the kind color are the opposites penetrating/compressing, it is superfluous to posit the pale/dark pair as additional principles for that kind. In this kind of case, although there are two oppositions for a substratum, only one of the two oppositions is *a principle.* Once a primary opposition is identified, it is excessive, or going too far, to make the other a principle.

This reading requires that only one of the two oppositions is primary. This gains further support from the argument in the final passage of this phase:

> At the same time, it is also impossible for there to be many primary oppositions. (189b22–3)

Among other things, this would show that were there two oppositions for a genus or kind, it is impossible for them both to be primary oppositions. One of them would be superfluous (as a principle for the kind), and only the primary opposition is a principle for the kind. This would then secure the claim that for a given kind there cannot be more than two opposites *as principles.* Although this allows there to be a pair of oppositions for a single kind, the members of one of the oppositions do not have the status of principles for the kind as such.

The argument for the claim that there cannot be a plurality of primary oppositions serves as a clarification and refinement of the arguments in the opening phase, at 189a13–14 and 17–20, and proceeds as follows:

> For substance is some single genus of what is, so that the principles will differ from one another only in being prior and posterior, but not in genus; for in every case there is a single opposition in the case of a single genus, and all oppositions are thought to be led up to a single <opposition>. (189b23–7)

The premises that there is just one opposition for one kind, and that substance is a single kind, jointly support the claim that there is just one

opposition, hot/cold, is not primary. Simplicius grants that such a reading would have had some consistency had Aristotle actually said "were generated from each other" (ἐξ ἀλλήλων γεννᾶσθαι). But he objects that in fact Aristotle said that they are able to generate from each other, which Simplicius takes as meaning that both oppositions generate the same things. However, to say that the two oppositions generate ἐξ ἀλλήλων is not to say that they *independently* generate X. Rather, each opposition has a role to play in the generation of X, and they generate X from each other in this way: the primary opposition generates X by generating a secondary opposition that mediates the causality of the former and thereby proximately generates X. This is not a mutual generation of the two oppositions from one another, but rather the way in which the two oppositions mutually generate X.

opposition for the category of substance. However, one might challenge the assumption that a single kind has just a single opposition since there is a way in which this is not so for the kind color. However, in that case there still is just one *primary* opposition, and only the primary opposition counts as a principle. It is part of the purpose of the final phase of I.6 to make this clear.

We are told once again that substance is a single kind, but now with the additional clarification that it is a single *kind of being*. There are all sorts of kinds, with different oppositions associated specifically with different kinds, but when investigating the number of principles in general we need to consider the question at the maximal level of generality. Although considerations about the opposites for lower kinds help us with this goal, it is important to see that when he limits the number of principles to two or three, this limitation is to be understood accordingly. He is not claiming that there are not distinct principles associated with distinct lower kinds.

Since there might turn out to be more than one opposition for a single kind, the proposition that there is one opposition for one kind stands in need of a more accurate formulation. For instance, one way in which a kind might have two oppositions is illustrated by a theory of color that posits two oppositions for that kind. Likewise, a theory of being might allow a plurality of different oppositions that are principles for being. In that case, however, only one would be *primary*. According to this passage opposites for the kind differ only in being prior and posterior, but do not differ in kind. This last clause tells us that the difference between prior and posterior principles is not one of kind. One way in which this would be the case is if the principles did not even belong to a kind. The primary opposition for the kind color is not itself a member of that kind. Perhaps it belongs to some other kind, but perhaps it simply belongs to no kind at all. In any case, color is not one of the kinds of being, and its primary opposition is not the primary opposition for a kind of being. However, color is included in a kind of being, namely the category of quality, and the primary opposition for this kind of being would fall outside the kind. The primary opposition for the category of being is prior to the primary opposition for the sub-kind color. However, if at least one of these oppositions does not belong to a kind, then although there is a relationship of priority and posteriority, they could not differ in kind.

More generally, the primary opposition for each category fails to belong to a kind. Simplicius too took this final passage to commit itself to there being a primary opposition for each category of being, according to which there will be an opposition of form and privation for quality, another for quantity, and so on. However, he also assumes that the opposition of form and privation in the category of substance is the primary opposition, and that all of the other form/privation oppositions are in some way "reduced" to it. So, not only is the form/privation opposition for the category of substance prior to the oppositions for the sub-kinds within that category, but it is also prior to the primary opposition for each category. He associates the priority of the primary opposition for substance with his claim that the other oppositions are reduced to it. Its priority may even consist in the reducibility thesis. Simplicius thought that this reducibility followed the same pattern as the *pros hen* analysis of being that explains each non-substantial kind of being in terms of its connection with the primary kind of being, that of substance. The non-substance genera of being derive their being from substance, and he claims that in the same way, the being of the primary oppositions for the other categories is derived from and depends upon the being of the primary opposition for substance. However, this cannot be right unless the primary form/privation oppositions are themselves beings, which they cannot be since they do not fall within any kind of being.

A more defensible account of the reduction of all oppositions is needed. Since form and privation are not, when spoken of at the level of a categorical kind of being, themselves beings or entities in the ontology, they are not entities with natures that can be defined. Rather, for each and every change of a natural object there will be some specific kind of substance that undergoes the change, and there will also be some specific kind K with respect to which the substance changes. There will be a primary opposition for kind K, and the members of that opposition are related to each other as form and privation. The only entities capable of retaining their identity while receiving opposites are substances, and so the most general level of description possible for anything that changes is that it is a substance, and hence a substratum for properties that is not itself a property. Furthermore, any change of a substance involves determinate properties, and the properties that it is capable of receiving or losing have principles. Not only do they have subjects upon which they depend, and hence have a substratum of some type or other as a principle, but they also have principles of a different type. These latter principles are not substances, but rather the primary oppositions that are unique to the specific kind K with respect to which the subject changes.

At the most general level, the members of such a primary opposition are a form and correlative privation. Hence form and privation are not the specific principles involved in any specific kind of change, but rather for each kind of change there is a pair of opposites that are related as a form and its privation. What all of these specific form/privation pairs have in common is simply that they play analogous roles. As he argues in *Metaph.* Λ 4, "in one way the causes and principles of different things are different, but in another way should one speak generally and by analogy, they are the same for all."[13] What this suggests is that the reduction to which Aristotle refers at 189b27 is not a reduction of being, nor is it a definitional reduction, but rather involves characterizing their roles in the most general terms possible. There are three principles in each case, and by analogy they are the same. When looked at in this way, the primary form/privation opposition for substance is analogous to the primary opposition for each of the other categories of being, but these other oppositions are not reduced to it.

To say that all other oppositions are led up to a single opposition does not mean that they must be reduced to the primary opposition for the primary category. Each opposition is one and the same by analogy as any other opposition. To give a general account of the principles common to all change requires that we are able to say what it is that all specific principles for specific kinds of change have in common, and when we lead our investigation to this level of generality it will turn out that the commonality consists in an analogical unity. However, I.6 is not yet considering these questions of analogical unity, but rather is pointing to the need to identify the opposites that are principles and elements of all things, and this will turn out to be form and privation.

[13] 1070a31–3.

Physics I.7

David Charles

> This then is what we ourselves say, beginning with a discussion of coming to be in its entirety. It is natural first to speak of the common features and on this basis to examine what is distinctive about particular cases. (189b30–2)

Aristotle's seminal discussion of coming to be and related topics can be divided into several sections. Since the chapter is relatively long, I shall not consider all parts in equal depth but focus on those that introduce and develop his account of what underlies.

I. Initial Comment: What we Say about Coming to Be: 189b32–190a13

> We say that one thing comes to be from another and that something comes to be from something different, when speaking of simples or compounds. What I mean is this ... (189b32–3)

Aristotle gives several examples to indicate what he means, focusing on the case of a human who is not educated and becomes educated.

(a) A human comes to be educated.
(b) Something non-educated comes to be educated.
(c) A human that is non-educated comes to be a human that is educated.

Aristotle takes these three expressions ('a human', 'non-educated' and 'educated') to be about simples, while the 'A human that is educated'

This chapter, initially presented as a talk at Sapienza and Rutgers Universities in autumn 2013, was re-written in July 2014 and (in a slightly revised form) discussed in Sapienza in June 2015. I should like to thank the participants at those meetings for their helpful comments. I am also indebted to István Bodnár, Alan Code, Lindsay Judson, Sean Kelsey, Emily Kress, Harvey Lederman, Michael Peramatzis, Diana Quarantotto and Alex Zhang for further discussion of these issues. Thanks also to Huw Charles for insightful and precise comments on the final version.

and 'A human that is non-educated' are about complexes. He compares these with another expression used to describe the same case:

(d) From something non-educated, something educated comes to be,

which can be employed instead of (b). He notes, however, that we cannot replace (a) with

(e) From a human, something educated comes to be (190a5–8).

In considering these examples, Aristotle notes:

> Of the things that come to be, in cases we describe as involving simples, in one case something comes to be by remaining, in the other by not remaining. (190a9–10)

His remark can be spelled out as follows: in (a) a human remains from the outset and is still a human when he (or she) becomes educated, but in (b) and (d) neither being educated nor being non-educated remain through the change. Nor, he adds, is the situation altered with regard to being educated and being non-educated when these terms are compounded, as in (c). In none of these examples does being educated or being non-educated remain through the change.

Aristotle's initial discussion introduces the following point: in the cases of coming to be which he has discussed, one type of simple remains (*hypomenei*) through the change (as a human does in (a)) while other simples (such as being educated and being non-educated) do not. Further, the compound (a human that is educated/non-educated) does not remain through the change. On the basis of these remarks, he formulates a more general thesis.

II. First Thesis: A Two-Headed Claim

> In the light of these distinctions, one can grasp in all cases of things that come to be, if one looks closely at the phenomena in the manner in which we speak of them, that [I] something must always underlie (viz.: that which is in the process of coming to be) and [II] that this [that which is in the process of coming to be], even if it is one in number, is not one in form (by 'in form' I mean 'in account'). (190a13–16)[1]

[1] In this translation 'something' is the subject of 'underlies' with 'that which is in the process of coming to be' in apposition. An alternative (consistent with the general interpretation suggested) would be to take the latter as the subject of 'underlies' with something as the object (cf. 190a34). So understood, the clause would mean: that [I] that which is in the process of coming to be always

Aristotle's first thesis has two parts:

[I] in all cases where things come to be, there is (in each case) something that underlies; and
[II] that which is coming to be, even if one in number, is more than one in form.

Aristotle notes that in his example a human remains (*hypomenei*), while being non-educated and the compound (a human who is not educated) do not: 190a18. The human in question remains through the process of coming to be: first he (or she) is non-educated and then he (or she) is educated. Aristotle claims that one can see this structure in all cases of coming to be. There is always something that remains throughout the coming to be, which at the outset has one feature (such as being non-musical) and comes to have another feature (such as being musical).

There are several points to note about his thesis:

(i) Aristotle is specifying the relevant use of the term 'to underlie' (*hypokeisthai*: 190a15) on the basis of his description of cases in which something is said to remain (*hypomenei*). He makes this clear by invoking the idea of its remaining in the next sentence (190a17–18). He is not proposing, as an independent thesis, that what underlies in fact remains (or survives) through the coming to be. Instead, he introduces his conception of what it is for something to underlie (in the manner relevant to [I]) on the basis of the idea of something remaining throughout the coming to be. In his view, that which is one in number [II] remains throughout the coming to be. [I] and [II], so understood, capture his understanding of what it is to underlie as given by the examples used in 189b32ff. We are not offered, in the present context, a general idea of what it is to underlie independent of these examples. [I] and [II] tie the relevant use of the term 'to underlie' (and so of the associated terms 'underlying' and 'the underlier') to what remains, one in number, throughout the coming to be.[2] Aristotle is not, it seems, relying on the more theoretically rich conception of the underlier (*to hypokeimenon*) he develops elsewhere (such as in *Metaph.* Z 3.1028b35ff., where he identifies it with 'what everything else is said of and is not said of anything else').

underlies something and [II] that this (that which is in the process of coming to be), even if one in number, is not one in form.

[2] While Aristotle has talked of 'what underlies' in previous chapters (e.g.: 187a13), the present passage constitutes his attempt to specify more precisely what this term signifies in the context of coming to be.

(ii) Aristotle is making a claim about what we will see in all cases of coming to be, if we look at them in the way we (or, perhaps, some favoured group of us) talk about the examples just discussed. By means of these, we are introduced to the thought that there is in each process of coming to be some one thing that underlies, remaining throughout the process. While that which underlies may differ from case to case, there is a unified type of thing to which each individual underlier belongs (Aristotle's discussion might be taken as an account of what the general term 'the underlier' signifies: the type of thing that underlies). He claims that there is, in every case of coming to be, an instance of this type. The latter claim belongs to the second stage of scientific investigation (as described in *APo* II.10). In the first stage, one specifies what type is meant by the term 'the underlier', in the second one grasps that there is such a type. Only then can one proceed to a third stage of enquiry in which one searches for the underlier's basic nature or essence. To establish that there is indeed such a type, Aristotle needs to show that there is a properly unified type of thing – the underlier – present in all cases of coming to be. There is, after all, the possibility that this type, like pride (to use his example in *APo* 97b13–21), might dissolve into several quite distinct types of phenomena under more careful investigation. Or perhaps the differing cases are related focally or by analogy.[3] At the third stage of enquiry, a more revealing, and more fundamental, account of the relevant type ('the underlier') – perhaps in terms of matter or potentiality – will be required.

Aristotle's methodology is cautious. He does not commit himself to any of the following general claims:

(a) The fact that we (or some favoured group of us) speak in a given way shows (or even gives us good evidence) that something is the case; or
(b) The fact that we (or some favoured group of us) speak in a given way is best explained by something being the case; or
(c) The fact that we (or some favoured group of us) speak in a given way makes it true that something is the case.

[3] In *Metaph.* Z 13, Aristotle distinguishes two cases: in one a particular subject underlies its properties, in the other matter underlies actuality (1038b4–6). For a similar distinction between cases of what underlies, see *Metaph.* H 7.1049a24ff. It lies outside the scope of this chapter to consider how these differing cases are related.

Nor does he claim (at least at this point) that *all* the relevant linguistic data are consistent with his claim. His contention is simply this: if we look at all cases of coming to be in the way suggested in 190a9–10, we will grasp what is the case. The specific linguistic evidence he assembles suggests a way of looking at the phenomenon that reveals (if we look carefully) what is the case: something underlies the change. What makes the latter claim true is not how we speak but rather what we will see if we look at the phenomena in the manner suggested. The linguistic evidence provides nothing more than a useful clue which enables us to see, after further scrutiny of the relevant cases, what is involved in any case of coming to be.

In fact, as Aristotle is quick to point out, the linguistic data are more complicated than so far indicated. Further work is needed to confirm his initial thesis and to show that it applies to all cases of coming to be.

III. A Complexity in the Linguistic Data: A Refinement of (d) and (e) and a Possible Problem: 190a21–31

Aristotle notes a complexity in the way we speak:

> The expression 'this comes to be from that' (as opposed to the expression 'this comes to be that') is more used in the case of what does not remain, as when we say that what is educated comes to be from what is uneducated but do not say that it [i.e. the educated] comes to be from human. Nonetheless we do also sometimes speak in a similar way [i.e. using the 'this comes to be from that . . .' way] about what remains, saying that from bronze a statue comes to be, not the bronze comes to be a statue. However, we do speak both ways in the case of something opposed that does not remain: we say both that this comes to be from that and this comes to be that.[4] (190a21–8)

[4] There are several points of detail to note about my translation/interpretation of these lines: (a) 'Is more used' (190a22) is taken to mean that the expression 'this (the musical) comes to be from that (the unmusical)' is used more widely (and more appropriately) than 'this (the unmusical) comes to be that (the musical)' in cases where the referent of 'this' (in the second sentence: the unmusical) does not remain through the change than in cases where it does. (b) The lines 190a21–3 are not taken as meaning that the locution 'This comes to be that' is never used (appropriately) in the case of what does not remain. Indeed, in 189b35 it appears that it is (as in 190a27). The claim can only be that although 'this comes to be that' can be (and is) used in any case of what does not remain, we do nonetheless have a (reasonable) preference (in this case) for the locution 'This comes to be from that'. (I am inclined to take 'we' to refer, following a suggestion of Michael Frede's, to what well-trained philosophers would say. It need not indicate a view of what any competent Greek speaker would say.) (c) 'We speak in a similar way also in some cases about what remains' (190a24–5) is taken to mean: we use the expression 'This comes to be from that' in some such cases rather than the locution 'This comes to be that'. This leaves open the question of the permissibility of the latter locution in these cases. (d) The absence of an article in 190a25 before 'bronze' in contrast to the presence of an

These remarks require scrutiny. With reference to what does not remain (such as the unmusical specified above), Aristotle claims that we more often say (and hence prefer to say)

(d) 'From this, that comes to be'

than

(e) 'This comes to be that'

although (as he notes in 190a26–8) both locutions are permissible. While neither says anything false or nonsensical, we prefer to say (d) than (e).

There is a similar pattern in some cases of what remains: we prefer to say (or say more often):

(d:i) 'From this, that comes to be' rather than
(e:i) 'This comes to be that'.

For example, we prefer to say 'From bronze (S^1), the statue (S^2) comes to be', not 'The bronze (S^1) comes to be a statue (S^2)'.

Aristotle is not, it seems, saying that (e:i) is straightforwardly impermissible; only that we prefer (and may reasonably prefer) to use (d:i).[5] (e:i), for all that has been said so far, might turn out to be permissible. Nor, so understood, is he saying that (e:i) is straightforwardly permissible as some have suggested translating this sentence (with considerable ingenuity) as follows: 'we say that the statue comes to be from bronze, not *only* that the bronze comes to be a statue' (introducing 'only' into their paraphrase of the text to maintain philosophical plausibility).[6] Questions of the permissibility of (e:i), and of the ontological structure it suggests, are left open at this point.

In sum: Aristotle notes that we use two different locutions in talking about coming to be. One is of the form: 'This comes to be that', the other 'From this that comes to be'. These locutions may be presented schematically, with 'A' representing being educated, 'non-A' being uneducated, 'S' a human, 'S^1' bronze and 'S^2' a statue. By '*' I shall signify something which (according to Aristotle) is not our preferred locution:

article before 'bronze' in 190a26 is not taken to point to different subjects: bronze (e.g. as a stuff) in a25, this piece of bronze in 190a26 (as Harvey Lederman (2014) has suggested). The absence of the first article is understood as stylistic, not as pointing to a change in referent.

[5] Jones 1974. [6] Code 1976a. His suggestion is adopted by Bostock 2006a: 4 n. 6.

[A] *'x comes to be y. . .'*	[B] *'From x, y comes to be. . .'*
*(i) A, non-A	(i) A, non-A
*(ii) non-A, A	(ii) non-A, A
*(iii) S^1, S^2	(iii) S^1, S^2
(iv) S that is non-A, S that is A	(iv) S that is non-A, S that is A
(v) S, A	*(v) S, A

While the [A] locution is permissible (although not preferred) in cases (i) and (ii), Aristotle (in this passage) does not rule on its permissibility in (iii). If it turns out to be permissible, the [A] and [B] locutions will differ (with regard to permissibility) only with respect to case (v). If so, the [A] locution will apply to all cases of coming to be, even if the [B] locution is preferable in cases (i), (ii) and (iii).

Aristotle focuses on the complexity of the linguistic data but does not provide an explanation of the differences noted. One suggestion would be that the [B] locution points to the origin or starting point of the process of coming to be. If so, the infelicity of [B] (v) will arise because the human (S) is not the starting point of the change from which being musical (A) comes to be. In locution [A], by contrast, the basic idea is of one thing remaining as the underlier with different forms at different times. If so, [A] (iii) will be infelicitous (to the extent that it is) simply because it does not explicitly refer to a subject (S) which underlies the change (contrast [A] (iv) and (v)) or contain an obvious space for an underlying subject of predication (contrast [A] (i) and (ii)). It is certainly true that [A] (iii) does not appear, at first sight, to be of the form introduced at the beginning of the discussion. There is no subject (whether explicitly present or presupposed) which remains and is initially S^1 and subsequently S^2. But what follows from this observation? Perhaps some, but not all, cases of coming to be follow the pattern suggested by locution [A]? Or maybe there are distinct types of coming to be, some correctly described by locution [A], others by locution [B].

Aristotle leaves this issue open. Perhaps [A] (iii) describes a situation in which there is one underlying thing that has to be referred to if we are to capture what is going on. If so, the phrase 'the bronze becomes the statue' will (if correctly understood) be permissible, even if it is not what we prefer to say. Perhaps every case of coming to be not only involves the ontological structure initially suggested by locution [A] (as captured in the original double-headed thesis) but can also permissibly be described using the [A] locution. Or does our preference against [A] (iii) actually show that the [A] locution is impermissible in these cases?[7]

Aristotle's reply is to claim that, notwithstanding the complexity of the linguistic evidence, the ontological model suggested by locution [A] applies to all cases of coming to be. If this is correct, preference for locution [B] in case (iii) is not a good guide to what is really going on.

IV. Aristotle's Reply: Different Cases but the Same Structure: 190a32–b11

Aristotle begins his reply by separating two cases:

> In all cases other than <the coming to be of substance>, it is evident that there must be something underlying, that which comes to be. For when a thing comes to be of such and such quantity or quality or in a given relation, at a given time or place, something is always present underlying (since substance alone is not said of another underlying thing, while everything else is said of substance). But that even substances, and whatever things simply are, come to be from something underlying will be evident to one who looks closely. For always there is something which underlies, from which the thing generated comes to be: as in the case of plants and animals coming to be from a seed. (190a33–b5)

In the first set of cases (which involve quantity, quality or relation) it is clear that, as suggested by locution [A], something underlies, namely that which is coming to be something else. Aristotle claims that the ontological model suggested by locution [A] captures cases where substances come to be. Here too there is always something that underlies. What are his grounds for this claim?

Aristotle comments: 'always there is something which underlies from which the thing generated comes to be: as in the case of plants and animals coming to be from a seed' (190b3–5). This remark may seem surprising. It appears to re-define 'something which underlies' in terms of the origin of change: the seed from which the plant comes to be. If this is correct, Aristotle is now abandoning the definitional connection between what underlies and what remains throughout the change (which he introduced in his initial discussion of the cases covered by the [A] locution). The seed does not, after all, remain throughout the coming to be of the plant or

7 Does Aristotle take all pieces of linguistic usage to reveal the ontology involved? Perhaps, in his view, only some do. If so, he would need to determine – in the light of his favoured ontology – which linguistic usages offer a correct picture of reality.

animal. There is no actual seed present at the end of the process (e.g. in the fully grown plant).

However, it is not necessary to understand Aristotle's remark in this way. There is an alternative. In saying 'something which underlies from which the thing generated comes to be: as in the case of plants and animals coming to be from a seed' he may intend to claim only that there is some one thing that underlies the entire process from which the thing generated comes to be and that this is true in the case of plants and animals coming to be from seeds. At the initial point in the relevant process (the stage at which the coming to be starts), that which underlies is a seed but it need not remain so at all stages in the process. What remains through the whole process was initially a seed but need not be a seed at every subsequent stage. In terms Aristotle uses elsewhere:

> that which underlies is the same throughout [e.g. the change between opposites] although it is not the same in being. (*GC* I.3.319b3–4)

In Aristotle's example of a plant, that which underlies may be the same (in number) throughout the whole process of its coming to be, even though it is first a seed, next a sapling, etc. That which underlies, while remaining the same entity, will change in certain respects as the process develops. In any particular case, there will be a particular underlier of this general type: one thing which underlies change and remains the same, even though at differing times it may be different things: first an embryo, next a child, then an adult etc. All such particular underliers will be instances of one general type of thing: the underlier, with its own distinctive general nature. (I shall use such phrases as 'the underlier', whatever its nature may turn out to be, to refer to the type of thing of which all particular underliers are instances.) Aristotle uses similar terminology elsewhere in describing the present moment (the now): it too remains the same as time passes between 1.01 a.m. and 1.04 a.m., even though it differs in being (being first 1.01 a.m., then 1.02 a.m., then 1.03 a.m. etc.: *Phys.* 219b12–13, 27–8). At one time the present moment is 1.01, at the next 1.02. If the present moment is understood as an individual which is first 1.01 and then 1.02, it will closely resemble the particular underlier which is first a seed and then a sapling.[8]

[8] Some may prefer to see the present moment not as an enduring particular but as a type of entity of which 1.01 and 1.02 are instances. The relevant analogy, together with the more general issues it raises, requires further study. For some preliminary discussion, see Charles 2003. For a helpful discussion of Aristotle's relevant locutions, see Lederman 2014.

In the case of the coming to be of a particular oak tree, the underlier (so understood) will be the same in number throughout the whole process although it changes as the process develops. In Aristotle's terminology, that which underlies may differ 'in being' at different times (first being a seed, then being a sapling) while remaining the same: it is that thing (however its nature is finally to be understood at stage three) which is various things at various times.[9]

There are, it seems, two ways to understand the phrase 'always there is something which underlies from which the thing generated comes to be: as in the case of plants and animals coming to be from a seed' (190b3–5). In the first, 'that which underlies' refers to the origin of the coming to be, in the second to what remains throughout the process. Which is preferable?

(i) The second way maintains the original account of that which underlies introduced in 190a15. There Aristotle simply stated that what underlies remains, not taking this as a further step in an argument but rather (or so I suggested) as something implicit in the examples cited. (Indeed, if one subtracts the idea of what remains, the underlier, understood simply as the point of origin, could equally well be the initial complex: the unmusical human in the case he discusses in which the unmusical human becomes musical.) What is added at this point, according to the second interpretation, is a way to generalise his original idea by specifying the underlier not as a specific substance (such as a human) but (in the way just explicated) as that which underlies the process of coming to be. Aristotle has not, as yet, said anything about the nature of the entity in question or the way in which it constrains the process of coming to be. Instead, he appears to rely on the intuitive plausibility of the thought that there is one thing present throughout the transitions mentioned (such as those of the animal or plant from the seed, the statue from the clay, and the house from the bricks and stones). He has not yet suggested what will play the role of the sortal concept in these examples (a role comparable with that of human in his original case). That task belongs to a later stage of enquiry.[10]

[9] Compare the following case: a person on his (or her) way to becoming fifty may remain the same person throughout even though she (or he) was initially a foetus, then a child, then an adolescent ... While what underlies the process of becoming fifty remains from the outset when it was a foetus, it does not remain as a foetus throughout!

[10] Aristotle has more to say about that which underlies. He does not rest content with specifying the underlier simply as that which underlies the change. Instead he introduces richer, more theoretically

(ii) Aristotle is, in this passage, concerned to show that the original ontological model revealed by locution [A] (and captured by the second interpretation) applies to all the relevant cases. To do so, he needs to establish that there is always one (simple) subject which remains throughout the process of coming to be. However, if he now falls back on a different, and less demanding, definition of what underlies (as merely the origin of the coming to be), he will not have succeeded in his project. Indeed, he will have given it up. By contrast, on the second interpretation, he has generalised his original idea, relying on the more abstract idea of there always being an underlier present in any change (however the nature of the relevant type is to be understood at stage three of the investigation).

(iii) Aristotle's next remarks suggest that he is seeking to apply his original model to the cases at hand:

> Some of the things that come to be in the unqualified way come to be some by change of shape (as a statue), others by addition (as things that grow bigger), others by subtraction . . . others by composition . . . others by alteration (as things changed in accordance with their matter). It is evident that all things that come to be in this way come to be from what underlies. (190b5–8)

These comments are not, it seems, designed merely to support the truism that there is, in all these cases of coming to be, an origin from which the change starts. Instead they suggest that there is some one thing (that which underlies) present from the beginning which survives throughout the change in some modified way (either by being added to/changed in shape/growing etc.). It appears that the thing in question remains from the beginning, albeit changed, for example, in shape, quality, size or material features. What remains need not (indeed, will not) be a seed: it may come to lose the shape, size and many of the qualities of the seed.[11] It is enough that that which underlies (which is at various times a seed, a sapling etc.) remains the same. So understood, the examples cited in 190b5–8 play an essential part in supporting Aristotle's original claim that, if we look carefully, we will see that there is one thing which underlies – even in the problematic cases of the coming to be of substances: 190b1–3. Indeed, seeing what happens in this way is part of what is involved in looking carefully at the cases mentioned.

laden, concepts (e.g. of matter and of what is potentially a human) to spell out the basic nature of what underlies.

[11] The phrase 'the seed', so understood, is used to give a contingent way to fix the reference of the term 'the underlier', not to define what it is to be an underlier.

(iv) If Aristotle were to surrender his original thesis that what underlies remains throughout the coming to be, he would have no grounds for his next major claim: that every object that comes to be is complex (190b11–12). If he merely claims that each case of coming to be has a starting point, he cannot rule out the possibility that what results from such a process is a simple (non-complex) object: that is, one that has arisen from a starting point which no longer survives in the finished product. However, the idea of a continuant surviving from the beginning to the end of the coming to be appears essential for his next claim. (See section V below.)

On these grounds, it seems best to understand Aristotle's talk of 'that which underlies' (in 190b3) as referring to one thing which remains throughout the whole process. However, so understood, his argument raises several questions.

(1) Has he really shown that, in *every* case of coming to be, there is one thing which underlies that coming to be? He has certainly set out (in 190b1–9) a way of thinking about the cases he mentions which addresses the specific problem shown by the apparent infelicity of [A] (iii). However, he has not established that this way of thinking captures *all* cases in which substances come to be. Perhaps his concluding statement: 'all things that come to be *in this way* do so by . . .' (190b9–10) is intended to suggest that while his model applies to the cases just discussed, he has not shown that it holds universally. Or perhaps he is confident (but – if so – with what justification?) that the model just sketched can be generalised to cover all cases of coming to be, even if he has not shown that it does.

(2) Aristotle's description of his examples is compressed – indeed, telegrammatic. While the seed may grow into a plant, in what way does one thing remain underlying the whole process? As Aristotle himself remarks (in *Metaph.* H 7), the seed 'has to be placed in something else and change . . .' (1049a14–15). However, if the seed does not remain as a seed in the plant, in what way does that which underlies remain as one thing throughout the process of coming to be? How should we understand the key idea of the one thing that underlies?[12]

[12] Nor does Aristotle return in this chapter finally to rule on the apparent infelicity of [A] (iii). There are two possible suggestions, consistent with what has been said (so far): (a) this phrase is permissible provided that 'the seed' is understood to refer to the seed as that which at a given time is the underlier; (b) the phrase is impermissible as the seed, understood as the seed itself, does not become the oak tree. Perhaps, in the light of his discussion in 190b1–12, both answers can be seen as correct. If so, the infelicity of saying (without qualification) 'the seed becomes the oak' is that it does not clearly distinguish (a) from (b). Aristotle, however, chooses to focus on the ontological

It may be helpful, in considering both (1) and (2), to note that elsewhere, in theoretically more developed contexts, Aristotle uses phrases such as 'that which is potentially a . . . house' to characterise what is present throughout the process of the coming to be of a house (1049a8–11). That which is potentially a house is present when nothing needs to be added or changed for the process of house-building to get under way. This characterisation is consistent with the initial object being re-moulded (and changed) in the process of house-building, provided that there remains (at each step of the process) some one thing which is potentially a house. The seed will be potentially an oak if it is in a state in which, through the operation of the relevant internal causal principle (1049a15–16), it will become an oak. No further change in the seed (or, in the human case, the embryo) is required for this causal principle to begin its work. While what is potentially an oak was initially a seed, as growth continues it will be (as time goes by) a stripling, a half-grown tree etc. Indeed, as we shall soon see, what is potentially an oak may remain after the process of coming to be is completed (in the grown tree which fully realises the potential to be an oak).

Aristotle exhibits great resourcefulness in developing and defending his idea of what underlies in a variety of contexts. Nowhere is this clearer than in his discussion of elemental transformation in *De generatione et corruptione* I. There, it seems (or so I have argued elsewhere)[13] that what underlies any particular elemental transformation will always be something with the potential to become another element in the relevant cycle (although the thing in question will differ from change to change). There need not be one type of material substratum (such as water or fire) to which all such underliers belong. Instead, the relevant type can be defined as that which has the potential to undergo elemental change.[14] Aristotle's claim (so understood) is that, in each transformation, there is always one particular instance of this general type which remains the same throughout the change.

structure at work in this case and does not return explicitly to reconsider the linguistic difficulty (if such it is) he has isolated in [A] (iii).

[13] Charles 2003. In *GC* Aristotle frequently uses the phrase 'that which underlies' to refer to the type of object whose instances underlie particular changes, not to the individual instances of that type which underlie particular changes.

[14] But why think in terms of the same type of object at all? Does Aristotle need more than the relevant property 'having the potential to undergo elemental change' which is true of earth, fire etc.? Although this important issue (raised by Lewis 2008) requires further investigation, Aristotle seems reluctant to think of properties as ontologically basic, preferring to present them as belonging to a type of object which remains through the change. Indeed, he needs continuants of this type if he wishes to reject the idea that properties can (somehow) 'jump' from one object to another.

In *Phys.* I.7 Aristotle does not commit himself to the sophisticated positions he adopts elsewhere about the nature of that which underlies. His remarks leave open a range of possibilities as to how that should be understood. He does not even commit himself to the idea of the same 'matter' surviving through the process of coming to be. The underlier (whose existence the present discussion is designed to establish) is introduced in a more abstract way. While his project, as it subsequently develops, is to specify more fully what it is to be an underlier (in terms, for example, of matter or of what is potentially F), in *Phys.* I.7 he keeps his options open. The precise specification of the nature of the underlier will be established at a later stage in his enquiry. At present, Aristotle is merely pointing to the importance of that which underlies, leaving for further discussion the investigation of the basic nature of the type of entity involved. This may be why in *Phys.* I.7 he speaks of 'the underlying nature' (191a8) and 'the underlier' (191a19–20), while elsewhere – in more theoretically developed contexts – he talks of 'matter'.[15]

But why is Aristotle confident that there must always be something that underlies any process of coming to be? Does he merely rely on some inductive support for this claim based on the cases discussed? Or is he depending on a further more general argument? The present passage does not make this clear.

Aristotle's approach has considerable intuitive appeal if one thinks of the changes involved as ongoing processes. In the process of coming to be, it is natural to think that there has to be something present (at the first stage) in what is coming to be something that guides (or plays some role in guiding) the whole process, up to and including the final transformation into the finished product. If there were not, there would be nothing – in the object being changed – to constrain the development of the process or account for its final destination. What is present at the first stage must, if it is to continue to guide the process, be present (in some way) at each subsequent stage of the process (up to and including the final one). Otherwise there would be nothing remaining capable of constraining the process in the required way. Anything could be replaced by anything (provided perhaps that mass is conserved). There need be nothing present in the antecedent stages which constrains, by surviving in the appropriate way, what happens

[15] When the term 'countable matter' is used in 190b25 Aristotle is generalising (*holos*) over the examples he has given of what underlies, not using matter as a theoretical term. The use of 'matter' in 190b9–10 in talking of things being 'changed in accordance with their matter' looks to be a parenthetical, ordinary language, use of the term, not a theoretically loaded one. I discuss its use in 191b10 below.

next. If this is unacceptable, there has to be something that remains at each stage (including the final one) to ensure that what eventually comes to be is an F. Without this the process might fail at the last hurdle! If we talk in terms of potentiality, there has to be something with the relevant (passive) potential to be an F present throughout the process.

This line of thought does not depend simply on the need to distinguish between coming to be and replacement – where the former requires something to survive whereas the latter does not. It is based on the idea that, without something that plays the role of the underlier, there will be nothing to constrain the processes we observe (in the way we see that they are constrained). What is problematic about replacement, from this perspective, is not simply that it is different from coming to be (as Aristotle describes it). It is rather that it loses the idea of something present throughout the process (in the patient) playing a role in guiding it towards the final outcome (or, in some cases, the goal). For this, Aristotle may think, there has to be (*inter alia*) one thing which underlies, remaining to constrain the process of coming to be throughout.[16]

However, while this line of thought is intuitively appealing, it is not explicitly set out in *Physics* I.7. Is Aristotle presupposing an argument of this type? Or does he prefer, at the present stage of his enquiry, to introduce his basic claim (viz. that there is an underlier) in cases of coming to be without the theoretical support provided by his later account of processes, their subjects and causes? The latter approach might, of course, be appropriate if his goal in the present chapter is only to establish the existence of the underlier, not to provide a full account of its basic nature. (I shall return to this issue below.)

V. 190b10–17: Summary and the Next Move

> From what has been said it is clear that everything that comes to be is always a composite: there is something that comes to be and something that comes to be that <thing which comes to be>. There are two types of the latter: either that which underlies or what is opposed <to that which comes to be>. I say that the uneducated is opposed and human underlies; and shapelessness and formlessness and lack of order are opposed and bronze or stone or gold is what underlies. (190b10–17)

[16] There has to be an efficient cause at work also: see *Metaph.* Θ 7.1049a1–12 for a more detailed discussion of this issue. Both efficient cause and matter are required to account for the relevant changes.

In these lines, Aristotle makes a further move. If in all cases of coming to be, there is something that underlies and two opposed features (such as being A, being non-A), what comes to be will always be a compound in the sense that it contains (a) what underlies and (b) something opposed to the initial condition from which the coming to be began.

The 'always' in this claim might be understood in two ways. It might be taken to govern the whole career of the composite object, requiring that what comes into being as a composite entity always remains as a composite. However, Aristotle has so far only shown that it is always the case that what comes to be – at the time it comes to be – is a composite containing (in some way) that which underlies (whatever that may be) and a given 'opposite'. There is no requirement (as yet) that what comes into being as a composite must always be a composite. Perhaps in its subsequent development it could cease to be a composite and become non-composite as the two initial components develop into a simple unity. Nothing so far rules out this possibility.

Aristotle seems committed to a third claim (albeit one expressed with considerable caution):

> [III] It is always the case that what comes to be – at the point at which it comes to be – a complex entity containing (in some way) that which underlies and a given 'opposite'.

But is what comes to be essentially complex? Or is it only accidentally complex? Before we consider Aristotle's response to this question, it should be noted that in the present passage, as in earlier ones in this chapter, he speaks of things being opposed (*antikeimena*) rather than being opposites (*enantia*). He seems willing to move from talking of well-defined opposites (such as the hot and the cold) to talking about positive states and their privations. While this change in terminology may be important, I shall leave it aside in order to focus on what underlies. It will, in any event, be raised, even more acutely, by discussions of other passages of *Physics* I.

VI. 190b17–29: The Introduction of Principles: The Final Moves

> It is evident that if there are indeed causes and principles of things which are by nature – from which first causes and principles these things are and come to be non-incidentally what it is said to be in describing its nature (*ousia*), then everything comes to be both from what underlies and its shape. For an educated human is, in some way, composed of a human and educated, in that it is decomposable into the accounts of these. It is clear, therefore, that

> things that come to be do so from these. What underlies is one in number but two in form. The human, the gold and in general the matter which is countable <is one in number>. It is more a *this such* [i.e. a determinate particular] and the thing that comes to be does not come to be from it incidentally. The privation and the opposition is incidentally. The form is one: for example, the arrangement or being educated or any of the things predicated in this way. (190b23–9)

This dense passage contains a number of important claims (one put forward in the antecedent of a conditional). I shall mention three:

(a) The principles of change and of being for natural objects are the same (this is introduced in the antecedent of a conditional).
(b) Natural objects come to be non-incidentally from the principles of change.
(c) What underlies is one in number and (in some way) a *this such* (a determinate particular).

Aristotle has already presented a number of considerations relevant to (b) and (c). But now he makes a further move, suggesting (albeit without argument and in the antecedent of a conditional) that the principles of coming to be and of being for natural objects are the same (namely, (a)). Why believe this to be the case? And what are its implications? (a) introduces a further claim:

> [IV] The principles of coming into being and of being for natural objects are the same.[17]

However, Aristotle does not offer any reason at this point to accept [IV] and little guidance as to how to interpret it. Is he once again simply putting forward a hypothesis to be confirmed (or disconfirmed) by subsequent investigation? Either way, [IV] plays a major role in Aristotle's subsequent argument. From [IV], together with the suggestion that the principles of coming into being are that which underlies and the form, Aristotle infers:

> [V] The principles of being are that which underlies and the form.[18]

[17] Aristotle does not make explicit what 'sameness' requires in this context. Are the principles simply co-extensive or are they more closely connected (see *Metaph.* Γ 2.1003b22–5)? If they are more closely connected, are they numerically identical or (e.g.) inseparable in definition?

[18] On this view, the deprivation (*steresis*) will not be a principle of being or of coming to be. Although Aristotle may incline to this conclusion in 190b25–7, further argument in *Phys.* I.8 is needed to secure this result. In [V] 'principles' refers to what plays a given role (such as being an underlier or being a form), not to the role itself.

But what does [IV] mean? Why should we accept it?

It is tempting to re-construct Aristotle's pattern of thought as follows: the principles of change specify conditions necessary and sufficient for the coming to be of (say) an oak tree. As such, they must include the conditions for what is arrived at (at the end of the process) being an oak (See [III] above). But, if such conditions are necessary and sufficient for the presence of an oak (at the final stage of its coming to be), they will be ones whose continued presence is necessary and sufficient for the continued existence of an oak. Otherwise they would not be principles for the coming to be of an oak. So – one might conclude – the principles in question are (i) that which underlies and (ii) the form. Between them they will account for the coming to be and the continued existence of substances. There is one underlier which remains from the beginning of the coming to be to the end of the existence of the object in question. That which underlies is a particular continuant which persists from the beginning of the process of coming to be until the death of the oak.[19]

Perhaps this was how Aristotle was thinking at this point. But if it was, he was contemplating a major step without giving much justification in the immediate context. We can see this by considering two possible objections:

(1) There might be more to being a principle of being for an object than merely playing an important role in the object's coming to be. Perhaps such principles should explain the presence of all an object's necessary properties (and characteristic activities) as well as locating it in an appropriate definitional slot (in terms of its genus/*differentia* etc.). These requirements, suggested in *Posterior Analytics* (and developed elsewhere), constrain what it is to be a principle of being for an object. It is a major claim to suggest that these are the same principles as those that govern the process of becoming the object in question. Simply being necessary and sufficient for the existence of the object will not be enough. The principles of being should capture the essence of the object in question.

(2) The factors present throughout the continued existence of an object might be different from those involved in its coming to be. Perhaps some factors maintain an object (or condition) once it has come to be, others account for its initial coming to be. What is gained by 'lumping' these factors together as those which constitute, at different times, that which underlies the object in question? Is there a well-motivated idea of that

[19] For more on this way of thinking, see my discussion of Aristotle's account of the role of matter in *Metaph.* Η 7 and 8: Charles 2010: 168ff.

which underlies (a continuant) which applies both to what is present through the coming to be of an object and its continued existence?

It is difficult to address the first concern at this stage in Aristotle's discussion. The problem is not merely the lack of supporting argument. It is more basic. The very notion of a principle of being has not yet been clearly specified. Are such principles simply conditions necessary and sufficient for the existence of the object in question?[20] Or should they meet some further additional *Analytics*-style constraints on definition? Is Aristotle independently convinced that the defining principles for the being of a natural object are those that govern its coming to be? Are the principles of coming to be governed by the essence of the resulting object?

While these questions are pressing, they are not resolved within the confines of *Phys.* I.7. As a result, it is not obvious whether, in this context, Aristotle is contemplating a major step (with less than adequate defence) or a smaller one: simply understanding the relevant principles of being as whatever is necessary and sufficient for an object to be – without any reference to further definitional or explanatory roles that such principles might be expected to play. Perhaps his subsequent discussion in *Phys.* II.1 and beyond provides support for this major step. Indeed, perhaps it is only once he has developed his richer conception of matter as what is potentially the object in question (as in *Metaph.* H 7.1049a1ff.) that he can fully defend the major claim mentioned here – without support – in the antecedent of a conditional. (It is, it should be noted, one thing to think that the objects of physical investigation come into being and pass away, another to think that the principles that govern their coming to be also govern their continued existence.)

The second concern is also pressing. Has Aristotle succeeded in supporting his important claim that there is one continuing thing which survives remaining from the beginning of the object's coming to be until the end of its existence? Has he laid a secure foundation for an investigation into the type of entity in question?

One might challenge the latter claim as follows: why assume that what underlies ceases to be present when the object in question ceases to exist?

[20] In *Phys.* I.2.184b22ff. Aristotle remarks that those of his predecessors who considered the principles of existing things were, in effect, searching for their origins. However, it is far from clear that Aristotle himself is engaging in exactly the same project, with no further conditions on what it is to be a principle of being. Earlier in *Phys.* I.2.185a12–13 Aristotle does note that he is assuming (as a basic claim) that 'all or some of the things that are by nature are changing'. But it is still a major step to conclude from this that being subject to change is a *principle* of the being of natural entities (even of those that do change).

Why not think of it as remaining through the process of passing out of existence, underlying the transition from (say) being a horse to being a dead-horse? One might suppose that there is something which guides this transition (at least in many cases), accounting for the orderly phases of death, putrefaction and disintegration.[21] If there is something which underlies the process of coming to be, should there not also be something which underlies that of passing away? But if there is, there will be something that remains from the beginning of the process of coming to be a horse, throughout the period when it is a horse, and continues to exist during the process of its destruction into a dead-horse. Let us call this 'the sch-underlier', which governs not only the coming into being and continued existence of the object but also its subsequent destruction. If there is a sch-underlier, it will not be a principle of the being (or even of the potential being) of the object in question. In those cases where an object (like the horse) cannot once again come to be a horse after it has died, what sch-underlies will not be (in its nature) the principle of the being of a horse. It will have to be defined in terms independently of the being of the horse. (A dead horse is not a potential horse, however much it is flogged!) Had Aristotle thought in this way, his whole account of matter would have been radically different. Indeed, he might have succeeded in developing something closer to organic chemistry (as we now understand it). Many will think that – had he done so – he would have chosen a better road! So why in *Phys.* I.7 did he focus on the process of coming to be, when his predecessors had concerned themselves both with it and passing away? Did he have good reason to do so?

If Aristotle makes progress with this issue in this chapter, it will be when he discusses the notion of what underlies in 191a7–12. I shall consider that passage next (out of order) before turning back to his remarks on the number of principles.

VII. More on That Which Underlies: 191a7–12

> The underlying nature is knowable by analogy. For as bronze stands to a statue or wood to a bed or the matter – that is, what is shapeless before it takes on shape – stands to what is shaped, so the underlying nature stands to the substance, which is a this such [i.e. an informed object] and what is.

[21] Consider, for example, the way in which an apple grows from a seed but then ferments, becomes rotten, comes to be full of maggots and finally disintegrates. István Bodnár notes that Philoponus and Simplicius consider somewhat similar examples in their discussion of horse carcasses with reference to *Phys.* I.8.

> This then is one principle, although it is not one in the way a this such is one nor does it exist in the way a this such exists . . . (191a7–13)

If 'the underlying nature' describes the type of thing that underlies a substance, Aristotle is here attempting further to characterise that which underlies. While previously he had suggested that there is something that remains from the beginning of the process of coming to be into the finished product, he had said little about the type of thing this is. However, even if his present remarks are designed to address this gap, they are cautious and rest on an analogy:

AS what is shapeless before it takes on shape to what is enshaped
SO the underlier is to the composite (an informed determinate object)

Bronze and wood are his examples of what is shapeless (in the relevant sense), statues and beds his examples of informed substances. What is shapeless is what is to be (or could be) enshaped (it awaits shaping) while what is enshaped is the finished product: the bed, the result of what is to be enshaped undergoing the process of shaping. What is to be enshaped (the wood etc.) survives the process of enshaping and persists in the enshaped object. In these cases, what is shapeless plays two roles: it was what was there to be enshaped at the outset of the process and is present, actually enshaped, after the process of shaping has been completed. What is to be enshaped (e.g. the wood) remains from the outset, undergoes the shaping process, and is still present (having undergone the process of shaping) in the enshaped object. Bronze and wood play this role in the formation of statues and beds. The underlier stands to the composite as what is shapeless (and to be enshaped) stands to the enshaped object: it is what persists from the outset, undergoes whatever is involved in the formation of the composite substance and is still present in the composite object (the informed substance). Our understanding of the idea of the underlier present in the informed substance is based on the thought of there being something that remains from the outset through the process of coming to be (the shaping) and survives in the informed (enshaped) substance.

Why did Aristotle use this analogy? Since the idea of the underlier is more general than the three cases mentioned (bronze, wood, what is shapeless), he may be trying to characterise what it is to be an underlier in a way which covers a wide variety of cases. Further, since the idea of the underlier is more abstract than the particular material examples (bronze, wood, what is shapeless), he may be attempting to specify what it is to be an underlier in terms of its general role: as what remains through the process of coming to be and is still present in the composite substance. The underlier (understood as the type of thing that plays this role) cannot be grasped by perception (as wood, bronze and what is shapeless can). It can only be understood by reflection on the idea of there being something which remains in the way just specified. Indeed, the type of object in question may turn out (on further analysis) to be graspable only by analogy (as Aristotle remarks elsewhere of potentiality and actuality: *Metaph.* H 6.1048a30ff.).

What is added by these remarks? Aristotle, as earlier at 190a1–12, seems to introduce the general notion of what underlies in the case of substances by example. To be an underlier is to be what is present in a way similar to that indicated in the examples cited. Instead of offering a theoretical understanding of what it is to be an underlier, he gives a possible route to grasp (come to know) what it is to underlie (as is suggested by the term 'is knowable': 191a8). He does not yet provide a general account of what it is for something to remain as an (or one) underlier throughout the process of coming to be and the continued existence of the substance. The present passage simply points to a way in which we can latch on to the nature of the underlier.

Do we learn anything further about that which underlies? Much depends on the range of cases that are regarded as similar. In the cases explicitly mentioned, the wood and the bronze are what underlies. These, no doubt, remain the same through the process of generation of the statue and its continued existence. But why think that there is something similar in the case of the seed, let alone in elemental transformation or the generation of mixtures? One can, of course, fall back on the bare idea of that one thing (or type of thing) which underlies the coming to be and the continued existence of the object in question. But, since we have not been given a general understanding of what type of object this is, we lack a secure grasp of what needs to be present in cases of mixture or elemental transformation. Nor is it established that these cases are similar in relevant respects to the examples involving wood and bronze. Perhaps there is only an underlier of the type required in these specific cases!

Do Aristotle's remarks help to address our earlier problem of the sch-underlier? While he clearly focuses on the underlier not the sch-underlier – since he is concerned only with the process of coming to be and not with that of destruction – he offers no reason for his choice. The underlier is that which remains through the coming into being and continued existence of the object in question. But why deploy this idea and not that of the sch-underlier: that which remains throughout its coming to be, continued existence and passing out of being? In the present passage, Aristotle seems unaware of any such alternative, apparently blind to any but his own preferred way of thinking about what underlies.

Aristotle, of course, has the resources elsewhere, some derived from his teleological conception, some from his idea of matter as the potentiality to be an object, to defend his way of understanding the underlier. If, as he seems to think, there are only potentialities for positive states, there will be a potentiality to be a horse present from the outset which remains for just as long as the horse endures but no potentiality to be a dead-horse. This is perhaps why his developed theory concerns underliers not sch-underliers.[22] In its final version, the underlier itself ceases to exist when deprived of the form and the potential to become informed.

However, there remains a problem. In *Phys.* I.7, Aristotle appears to offer some general considerations designed to lead us to accept his own rich and elaborate theory of the nature of that which underlies. But if this is his aim, he should not now rely on that theory to support his initial account. Instead, he should show that there is something that exists as the underlier (at stage two of his enquiry) before going on to provide a fuller, stage three, account of its nature (e.g. in terms of matter understood as potentiality for form and his preferred form-directed teleology). If this is his strategy, he has to establish that there is such a thing as the underlier before he can construct his theory of what it is to underlie. Failure to do so will expose him to the charge of constructing a somewhat unmotivated theory of matter on the insecure basis of an undefended initial preference for the underlier over the sch-underlier.

A critic will see Aristotle's teleologically directed theory of matter as an attempt to build an elaborate theoretical edifice on the shaky foundations of the idea of the underlier, when he had no good reason (at this point of the argument) to take that rather than the sch-underlier as his starting point. Some defenders of Aristotle may prefer to see his account of matter (as a whole) as resting on considerations (drawn, for example, from his discussion of the principles of being of substances) that go far beyond those deployed in his account of coming to be in *Phys.* I.7. From their perspective, his discussion of coming to be and continued existence in this chapter

[22] In I.9 (192a9ff.) Aristotle considers and rejects a (possibly Platonic) proposal in which there is, it seems, one thing which underlies both coming into being and ceasing to exist. Some of his criticisms depend on his own teleological picture, others on his view of deprivation (*steresis*). For detailed discussion of these arguments, see James Lennox's essay in this volume.

is not the *foundation* for his subsequent theory. Instead, it is his theoretical commitments, subsequently revealed, that led him to focus here on the underlier not the sch-underlier. However, neither defender nor critic can construct a defensible unidirectional argument from a secure foundation laid in *Phys.* I.7 to the richer theory of matter advanced elsewhere.

It is important to note – before continuing – that, while Aristotle sometimes focuses on the underlier as a general principle (perhaps best captured by the idea of that which underlies – whatever its basic nature may be), he is happy throughout (190b24–6, 191a9–11) to drop a level and point to particular items that play this role: human, gold, wood, bronze and countable matter, understood as bits of matter (such as a lump of bronze, this piece of gold). For present purposes, it is sufficient to note that he can, without difficulty, talk of particular instances of what underlies while keeping his gaze firmly on the relevant type. Throughout he uses particular cases to exemplify his general claim about what it is to be an underlier. A similar pattern of thought is clear in his reference to plants and animals emerging from the seed (190b5) and in his later discussion of items involved in particular cases of coming to be (191a4–6). Particular cases (this matter/this seed/this thing being present) are used as examples of his general thesis about what it is to be an underlier.

VIII. Two Further Issues

[N] How many principles are there of coming to be?

This question, one might have thought, would have a straightforward answer in the light of Aristotle's remarks in the earlier parts of the chapter: there are three (two opposites and that which underlies). Indeed he seems to say as much at the very end of the chapter:

> But it is clear that there are three principles, in what way there are three, and what kind of principle each is. (191a20–2)

Along the route, he has considered the possibility that there are only two principles (the opposites: 191a15–16) but noted that we should add a further principle (that which underlies) which is not itself an opposite. Since these are, as he remarks, principles of different kinds (191a12–14), regarding that which underlies as a principle does not call into question the status of the opposites as principles. That which underlies, as he notes, must be present because without it opposites cannot act on each other (190b32–3). There has to be something present which was initially cold and becomes hot. Aristotle is, in this way, extending the range of principles beyond the traditional range of opposites. He does, however, consider a different answer:

> But, in another way, it is not necessary <to have three principles>: for one or other of the opposites will be enough to make the change by its absence or presence. (191a5–7)

Perhaps one does not need both opposites. One may be enough – if one considers one of the opposites as the privation of the feature used to define the object in question (191a12–13). One will then have a positive state (e.g. shape) which the underlier comes to have in certain conditions and the privation of that state (when neither the object nor the underlier continue to exist). The positive state, so understood, is (a) is required to make the underlier into a fully determinate object and (b) such that its subsequent loss marks the demise of the object (and, it will later emerge, the underlier) in question. It has to be defined in ways which essentially involve the underlier. The relevant positive state (such as shape) is that which the underlier (a) comes to have at the endpoint of the process of coming to be, (b) continues to have as long as the composite survives and (c) ceases to have when the composite passes out of being (at which point the underlier ceases to exist). Positive states of this kind essentially involve (and cannot be defined without reference to) that which underlies and the changes it undergoes.

As Aristotle's account develops, he introduces talk of form (and nature) to fill the logical space marked out by the 'shape' (or the relevant 'positive state') of the underlier. Given the connections between the positive state and the underlier, it is no surprise to find that physical form (as the further specification of the positive state) is essentially connected with matter and change, compared with snubness (a matter-involving form) and contrasted with mathematical form, which essentially involves neither matter (understood as an underlier) nor change (*Phys.* 192b1–2, 194a12ff.). However, although Aristotle's remarks in 191a5–7 may point in this direction, he has to take several important steps to arrive at his fully developed position. It is a non-trivial task for him (and for us) to spell out the required argument and resulting position.[23]

[Q] That which underlies: determinate object

[23] For further discussion of physical form as essentially matter-involving, see Charles 2008 and Peramatzis 2011. Snubness, we argue, should be seen as a material-way-of-being-concave (a distinctively matter-involving type of concavity) not as a combination of geometrical concavity (a matter-free geometrical property) and matter. These issues remain controversial.

That which underlies is described as countable and in some way like a determinate particular because what comes to be (the resulting determinate particular) comes to be from it (190b24–5). However, it need not be one in number in the way a determinate particular is (191b12–13). Elsewhere he talks of 'this wood' (presumably: this piece of wood: *Metaph.* Θ 7.1049a24), apparently distinguishing it from a determinate particular (a '*this such*': 1049a27–8) that is (in some way) predicated of matter (1049a35ff.). However, whatever Aristotle's final view may be, it is not set out in *Phys.* I.7, where he is content to point to an issue to which he will return. He does not yet have a sufficiently robust conception of a determinate particular to engage with the difficult question: in what way (if any) is this piece of wood not a (fully) determinate particular? Given his lack of specificity on this issue in *Phys.* 1.7, it will not be clear whether the form or that which underlies is the substance (*ousia*) (191a19–20). Indeed, if 'substance' indicates the nature of the object in question (as in 190b18), this issue recalls Aristotle's famous question in *Metaph.* Z 3: 'is the substance the essence ... or that which underlies?' (1029a34ff.). He will need more resources – precisely those provided in *Metaph.* Z 4–17 and beyond – to make significant headway with it. Once again, his present discussion points forward to the need for further work before his account is completed.

However, Aristotle's caution has a further consequence, directly relevant to our understanding of *Phys.* I.7. As noted above, when he considered the possibility that the principles of change and of being of natural objects are the same (in the antecedent of a conditional in 190b17ff.), he did not specify what is required to be a principle of being. If one takes the nature of the entity (its *ousia*) to be its principle of being, Aristotle's question at 191a19–20 indicates that he is not yet committed to that which underlies being a principle of being for an object. (Indeed, he may have been well advised to introduce his claim in the antecedent of a conditional!) Perhaps that which underlies is a principle of being (as well as of becoming) but not the nature (or essence) of the object in question. There are several possibilities left open at this point:

(a) That which underlies is a principle of becoming which survives in the finished object as its nature (*ousia*) and a principle of its being;
(b) That which underlies is a principle of becoming which survives in the finished object as a principle of its being but not as its nature (*ousia*);

(c) That which underlies is a principle of becoming which survives in the finished object but is not either a principle of its being or its nature (*ousia*).

In *Phys.* I.7 the notions of 'principle of being' and 'nature' (*ousia*) are not sufficiently determinate to establish which of these options is to be preferred.

IX. Some General Remarks and Questions

Aristotle, I have argued, makes three significant moves in *Phys.* 1.7:

[A] He introduces and partially elucidates the idea of that which underlies, understood as something which remains through the coming to be of certain objects.
[B] He argues that there is something which underlies in this way all cases of the coming to be (theses [I] and [II] above).
[C] He introduces the idea that what underlies is not only a principle of the coming to be of objects but also of their continued existence (theses [III], [IV] and [V] above): there is one continuing thing which remains from the beginning of the process of generation of an object to its final destruction.

He uses theses [I], [II] and [III], together with the linking claim in [IV] to arrive at a conception of the principles of being for the objects in question. While part of his aim in *Phys.* I.7 is to show how the claims made in I.5 and I.6 are true and coherent, he proceeds by developing a new line of thought which begins from a more careful account of coming to be than was given in the previous chapters and ends with a thesis concerning the principles of being.[24] While this line of thought contains several gaps and problems, it constitutes a new (and improved) route to grasp the principles of being.

Aristotle's introduction of that which underlies (as understood in theses [I]–[V]) is designed to locate something which he will later specify more fully using his ideas of matter (which as a general theoretical concept seems

[24] Sean Kelsey (2008) has correctly emphasised (i) the importance of seeing *Phys.* I.7 in the context of the partially aporetic preceding chapters and (ii) Aristotle's caution in introducing the notion of matter in this chapter. However, while Aristotle is concerned in this chapter to show how his earlier claims are correct, he does so (if I am correct) by spelling out a line of thought which indicates what the underlier is, how it is (or might be taken as) a principle of being and how it is connected with the other relevant principles (such as opposites or form and privation).

absent from the present chapter) and of potentiality. In doing so, he will deploy the richer conceptual resources needed to characterise, in a theoretically satisfying way, the type of entity he has marked out using his idea of that which underlies.

However, the theoretical innocence of *Phys.* I.7, which reflects the cautious, stage-by-stage fashion in which Aristotle constructs his position, has its own dangers. Several of his key ideas about that which underlies, although of considerable intuitive appeal, are not, or so I have suggested, articulated or defended with the detail required to make them fully convincing. There are, as already noted, major gaps in his presentation. For some, these show significant weaknesses in the foundations of Aristotle's further, more elaborate, theories of matter and potentiality.[25] Others, perhaps more charitably, will suggest that these gaps indicate only that the present chapter is intended merely to introduce a way of thinking that is to be justified fundamentally by the overall cogency and explanatory power of the resulting theory. From their perspective, several basic contentions in *Phys.* I.7 rest on claims which are established only in that theory. The chapter, so understood, is not Aristotle's attempt to argue for his overall theory of matter on the basis of generally accepted, theoretically innocent, premises. Instead, it should be seen as part of an early stage of his enquiry, designed to encourage us to think in one specific way about the issues at hand.

While it is no small matter to adjudicate between these two conflicting interpretations of Aristotle's project, such a task lies beyond the confines of a study of *Phys.* I.7 to do so. Much remains to be understood about Aristotle's method, aims and underlying epistemology in this book.

[25] For some *Phys.* I.7 provides the epistemological foundation for Aristotle's subsequent claims about matter; for others, it constitutes merely one helpful way to argue for those claims on the basis of general ('logical') grounds (which provide only defeasible support for claims made in the completed 'physical' theory). While the differences between these two viewpoints are important for an understanding of Aristotle's epistemology, they cannot be pursued here.

Physics I.8

István Bodnár

I

By the end of chapter 7 Aristotle was in a position to conclude:

> From these considerations it is now evident what distinguishes the opposites, how the principles are related to each other, and what type of thing that which underlies is. Whether the form or that which underlies is substance is not yet clear. But it is clear that there are three principles, in what way there are three, and what kind of principle each is. (191a17–21)

Accordingly, this closing section of the chapter suggests that the investigation of the principles of change has successfully been finished. This is then only reinforced by the last sentence of the chapter:

> This completes our examination of how many the principles are and what they are. (191a21–2)

All in all, one could take away the impression that the principal discussion of the book is over, and what remains is a discussion of the claims to substantiality of the two crucial principles, matter and form. But (though one may indeed argue that something along these lines will *also* happen in the following two chapters, which close Aristotle's argumentation in book I, especially in chapter 9) the last sentence of chapter 7 does not lead over to an explicit discussion of these issues about the substantial status of matter and form. Instead, in a typical Aristotelian manner, we have an introductory sentence attached closely to the concluding clause of the previous chapter. The usual *men* … *de* construction after claiming (in the first, *men* clause) that the direct discussion of the number of the principles, and what they are, has been concluded, leads over in a conjoint *de* clause to a further claim, that the solution canvassed in the previous chapters does also provide a resounding answer to the difficulty, or even puzzle, raised by some earlier philosophers.[1]

This puzzle has not been mentioned in the previous argumentation. At this point it is not pellucid why it is important to broach these issues at this place, and what is gained by showing that the Aristotelian account of the principles of change is a potent remedy – indeed, the only remedy – against this puzzle. One line of thought could be to suggest that chapter 8 gives further argumentative depth to the Eleatic position – a position that has been first discredited as not within the domain of physics, but then discussed nevertheless, because

> although <the Eleatics do> not <speak> about nature, they nonetheless happen to state physical difficulties. So it is presumably a good idea to have a little dialectical discussion about them. For the investigation does contain some philosophy. (185a17–20)

On this line of interpretation, even though the Eleatic puzzles have been laid to rest in chapters 2 and 3, chapter 8, by explaining how Eleatic monism is dependent on the Eleatic puzzles about generation from being or from not-being, from what is or from what is not, reaches two connected results. First, and perhaps most importantly, we are also provided with an account about why the Eleatics arrived at their monism: this happened when

> overplaying the immediate consequence [of the difficulty about coming to be] they say that the many things do not exist, but only what is itself exists. (191a31–3)

As a result of this, the original Eleatic puzzles of chapters 2 and 3 will be invested with additional relevance. Once the roots of the Eleatic error are here ferreted out by the Aristotelian account, it becomes clear how intimately these puzzles are, at their root, connected to an enquiry into natural philosophy.

However, we should note two further significant points here. One is that Aristotle claims that the resolution of these puzzles had to wait until his account of change has been thoroughly formulated. Although the Eleatic puzzle should have called for further inquiry, while later philosophers may have attempted to tackle some Eleatic problems, they could not deploy the Aristotelian explanatory framework developed in the

[1] Or to their perplexity. The word Aristotle uses is *aporia* – in the singular in the opening sentence of the chapter (191a23–4), in the plural in the closing lines of the chapter (191b30–4).

intervening chapters; instead they came up with suggestions that did not identify the root of the problem.

The other significant detail is that rather surprisingly Aristotle calls the Eleatics here 'the first to inquire philosophically into truth and the nature of the things' (191a24–5). This confers on them some crucial chronological priority. This chronological priority, however, need not situate them before every one of the natural philosophers (*physikoi*), who serve as the point of departure in chapter 4 for Aristotle's discussion leading up to his account of change. Rather, it suggests that the kind of rigorous philosophical analysis first advocated by the Eleatics – with rather disastrous immediate consequences – was also a crucially important starting point for formulating and assessing a viable account of coming to be.

II

We should note that Aristotle – unlike our contemporary account – does not present this Eleatic challenge as a watershed when he elaborates on the theories of change of his predecessors in *Physics* I. Most specifically, when he says about Anaxagoras, in *Phys.* I.4.187a26–b7, that his introduction of an infinite number of homogeneous and opposite principles was motivated on the one hand by the conviction that nothing comes to be from what is not, and on the other by the claim that opposites come to be from opposites, he presents the first conviction as the general, consensual opinion of the *physikoi*,[2] and does not connect it to Eleatic strictures on coming to be. The next step in Aristotle's account of Anaxagoras' reasoning from these two considerations is that the opposites which come to be have to be present in the entity undergoing the change (187a32). It is only after this interim conclusion that he reformulates these considerations in a dilemmatic fashion, that everything that comes to be has to come to be either from things that are, or from things that are not (both in the plural). The option that coming to be from things that are not is ruled out by all the *physikoi*, consequently only the other option remains, that everything that comes to be has to come to be from beings, and hence these have to be already – latently – present at the beginning of the process of coming to be (187a32–b1). Accordingly, despite the striking similarities, Aristotle does

[2] Indeed, not every one of the *physikoi* had to assert this general opinion in so many words. Aristotle could establish that all his predecessors upheld this principle by observing that each of them formulated a proposal in which some entity, or some entities serve as the starting point of the coming to be of other entities.

not formulate Anaxagoras' argument as a response to the Eleatic challenge. Rather, it is presented as following up on a consensual insight of all the *physikoi* and on their commitment that coming to be is from opposites.[3]

In contrast to this generally held opinion, in chapter 8 Aristotle introduces the Eleatic stricture against coming to be from what is not as formulated in an explicit manner, from appropriately general considerations – in particular, as one horn of a dilemma which is intended to block both possible avenues of coming to be. As I have indicated, Aristotle does not present the various accounts of his predecessors as a response to this challenge. His claim is that it remained to him to fully defuse this challenge. Earlier attempts, even if some may be registered, are found lacking, as only the fully developed Aristotelian account of change is adequate to this task. Plato, with his inadequate grasp of the material principle, is the only possible exception, because – unlike other earlier thinkers – he at least appreciated the inherent complexity of the underlier.

What this suggests is that Aristotle, instead of presenting the Eleatic challenge as a point of departure for later accounts, treats it here as a test case. If his account can meet this challenge, it is markedly superior to the other ones that cannot. In addition, Aristotle makes an even larger claim. In the opening sentence of the chapter he claims that the Eleatic puzzle can *only* be solved by relying on his account of coming to be and change. This is indeed a very large claim, and the ensuing discussion at first sight does not do justice to it. It is beyond doubt that Aristotle's argumentation intends to show that his theory of principles *can* provide a solution to these puzzles. But it is much more momentous to assert that any solution to

[3] It is also instructive to see what kind of responses to Eleatic challenges Aristotle mentions in the course of *Phys.* I. First there is mention of the 'more recent of the earliest thinkers' at the end of ch. 2, who went along with Eleatic worries so much that they initiated wholesale revisions of admissible philosophical language, because they were 'also troubled lest the same thing should turn out for them to be at the same time both one and many' (185b25–7). Apparently those like Empedocles and Anaxagoras, who 'say that the things that are are both one and many' (*Phys.* I.4.187a21–2), did not take the Eleatic menace to heart to the extent these 'more recent of the earliest thinkers' did. Chapter 3 mentions some other philosophers, who 'gave in to both [Eleatic] arguments, namely to the argument according to which all things are one, if being signifies one thing, by conceding that what is not is, and to the argument from dichotomy, by introducing atomic magnitudes' (187a1–3). Although ancient commentators take this to refer to Plato and Xenocrates, the Atomists may also be included, or may even have a better title for blocking *both* Eleatic arguments (for a discussion see Laura Castelli's balanced evaluation in section VI of her contribution in this volume). After this, when Aristotle claims at the beginning of chapter 5 (188a19–26) that every one of his predecessors posited opposites as principles he also includes Democritus, because he introduced the full as what is and the empty as what is not (with the superaddition of further oppositions within the genera of position, shape and order). So even if – or even though – Aristotle in ch. 3 took the Atomists to respond to Eleatic arguments, here he emphasises that consideration which links them to all the earliest philosophers (with Parmenides himself also included: see 188a19–22).

these puzzles *will have to* deploy the Aristotelian account. At this point I only intend to register this caveat. I do think there is a case to be made in defence of Aristotle's claim, and I will be able to make some brief suggestions in that regard in the closing section of my discussion.

In order to show that his account can address the Eleatic puzzle about coming to be, Aristotle first proceeds to set out this difficulty more or less on the Eleatics' terms:

> So they say that none of the things that are either comes to be or ceases to be, because it is necessary that what comes to be does so either from what is or from what is not; but from both of these it is impossible. For what is cannot come to be (because it already is), and from what is not nothing could come to be (because something has to underlie). (191a27–31)

The result of a putative coming to be in the first sentence is designated in two different ways. On the second occasion this is effected in a direct fashion, as 'what comes to be', the *gignomenon*, whereas in the initial formulation of the Eleatic thesis Aristotle speaks in the plural about the 'things that are', the *onta*: none of these beings comes to be or ceases to be. Nevertheless, neither Aristotle nor his Eleatic forebears should be worried about this terminological variation. Neither formulation carries any particular commitment; instead, in both formulations what we get amounts to setting out the requirements of a putative coming to be. Such a result of some coming to be should be a *gignomenon*,[4] because it is the result of the very process of this coming to be; moreover, if there are such comings to be, unless there is some fundamental ontological divide involved, the results of these should be full-fledged denizens of the world, they should also qualify as things that are, as *onta*. This introduction of the position in these terms is followed in the next sentence by a very succinct formulation, this time with the subject phrase 'what is' in the singular.[5]

More important than these considerations about the formulations on the subject term side are the actual reasons Aristotle sets out as deployed by the Eleatics. The considerations are formulated only for the impossibility of coming to be: the impossibility of ceasing to be is not argued for.[6] The rejection of the coming to be of what is happens in a dilemmatic fashion.

[4] Cf. Aristotle's meticulous distinction at the beginning of ch. 7 between the *gignomenon*, the item that undergoes change, and *ho gignetai*, the result of the change. This sharp terminological distinction, however, is given up at 190b3ff., where *to gignomenon* already refers to what comes into being from the seed or the semen.

[5] This formulation in the singular is also a vehicle of expressing generality. Moreover, switching to the singular fits better the monistic convictions the argument will give rise to.

[6] Cf. Parmenides B8, where also only the case for the rejection of generation is set out.

Either it should come to be from what is – but then we do not have a coming to be of what is, only its preservation – or it should come to be from what is not. This latter option, however, is impossible, because, as Aristotle says, 'something has to underlie'.[7]

III

It should come as no surprise that Aristotle's strategy with this difficulty will be one of introducing a distinction which will allow him to defend the coming to be of what is. Nevertheless, we should keep in mind at least two issues while we are pursuing Aristotle's solution to this Eleatic puzzle. One should be to elaborate on Aristotle's admissions: what are those cases in which he thinks that the application of his distinction makes the Eleatic claims come out true, and what reasons does he provide for evaluating these cases this way. The other one is how exactly the distinction is related to the account of principles Aristotle introduced in the previous chapters.

The distinction Aristotle introduces is one of something's being the case about a subject either in its own right, or something's being the case about the subject not in its own right.[8] The first of these is that in which we can make the assertion about the subject in the most proper way; the second of these – that in which something is the case about a subject not in its own right – also gives rise to perfectly legitimate and true sentences. Indeed, in most cases of the latter sort there must be some other way of specifying the subject. This in turn is made possible by the ontological complexity of the item picked out by the subject term of the assertion. In this case the entity corresponding to this further specification is

[7] Note that this paraphrase is meant to capture the ambiguity between the text in Simplicius' quotes in his criticism of Philoponus at the beginning of the commentary on book VIII, at *In Phys.* 1140.24 and 1144.7 – that it is the conviction of these thinkers, as recapitulated by Aristotle, that something has to underlie – and that of the codices of the *Physics* and of the commentators, including Simplicius' lemma at *In Phys.* 235.10 – that this is Aristotle's consideration, which somehow nevertheless can be used as an explanation of the stance of these thinkers. Either way, Aristotle's tinkering with the Eleatic position is unmistakable, and amounts to roughly the same: the Eleatic rejection of what is not, which guarantees that it cannot be the starting point for a putative process, is reformulated by Aristotle as an implicit admission that an underlier is necessary in such a putative process of generation, and what is not cannot fill this bill.

[8] In my paraphrase 'in its own right' stands for Aristotle's formulations with *hei* in the chapter, that something is true of an item in so far as it is what it is, and also in claims where something is true of an item in a non-coincidental manner (*me kata symbebekos*, or *ou kata symbebekos* if the negation is on the sentence level). 'According to itself' is reserved for those cases where Aristotle uses the expression *kath' hauto* in the chapter, or when I want to indicate that the restriction needs to be construed with narrow scope.

responsible for what was asserted in our original statement about the original subject truly, but not in its own right.

All this is achieved in three consecutive steps. First Aristotle insists that a distinction is to be made about the sentences formulating the Eleatic claims. As a next step he suggests how that distinction can be applied in a general fashion to any sentence, be it about any particular subject – about a physician, or about being. Then, returning to the claims about what is and what is not, he stresses that these claims will need to be assessed in the same way as the similar propositions about any other subject term.

> We, however, say that, in one way, to come to be from what is, or from what is not, or for what is not or what is to do something, or to be acted upon, or come to be a this, of any sort, is no different from the physician's doing something, or being acted upon, or something's being from the physician, or <something's> coming to be <from the physician>, and so as this <latter> is said in two ways, it is clear that also <being or coming to be> from what is, and what is doing something, or being acted upon, too <are said in two ways>. (191a33–b4)

These claims then are followed by the introduction of the crucial distinction:

> A physician builds a house, not in so far as he is a physician, but in so far as he is a housebuilder, and he comes to be pale, not in so far as he is a physician, but in so far as he is dark. On the other hand, he heals or fails to be a physician in so far as he is a physician. Since we say most properly that the physician does something, or is acted upon, or <something> comes to be from a physician, when these are acted upon him, or he does <these>, or comes to be <these> in so far as he is a physician, it is clear that also 'to come to be from what is not' signifies[9] '<to come to be from what is not> in so far as it is what is not'. (191b4–10)

After the introduction of the distinction Aristotle turns back to the Eleatics, and claims that they were unable to extricate themselves from their puzzle because they failed to make this distinction.

We should note that what Aristotle relies on at this point is not exactly the initial claim he made at the beginning of the chapter that the Eleatic error can be rooted out only through the results of the previous chapters. These are not even called in yet, at least not in an overt fashion.[10] Instead,

[9] Note that 'signify' here and at 191b26 is coordinate with what we say most properly, unlike e.g. on its occurrence in chapter 1, at 184b11. (Cf. further Laura Castelli's section on 'Signifying' (*semainein*) on pp. 87–91 above.)

[10] The contrast between something being that from which there is coming to be in its own right, or only coincidentally, was present in *Phys.* I.7.190b17–20, where coming to be is said to be from the

what he says that the Eleatics missed out on was a much more general distinction they should have made about what something is in its own right, and what it is coincidentally, not in its own right. Indeed, one could just retort that this should not be a workable distinction for an Eleatic, and what Aristotle is doing here is rerunning his previous strategy of arguing against Eleaticism by introducing some internal complexity into the unitary and uniform Eleatic being, and then using that internal complexity in order to argue for differentiation, and eventually for multiplicity.

But the Aristotelian charge here has some more depth to it. It is a two-stage argument. First the internal complexity is applied to the subject of coming to be and of perishing. The Eleatic may object that what could be the putative starting point of such processes does not allow for this internal differentiation, but then he fails to adhere to the structure of his own argument – as presented by Aristotle – where the unicity of being is derived as a second step, after the rejection of coming to be and ceasing to be.

Even though the distinction Aristotle uses here is apparently not introduced in terms of the results of the previous considerations, about underlier, privation and form, the way Aristotle opened this chapter suggested that there should be such a connection. This connection, or mapping, is effected in the next passage, in a context where Aristotle is not concerned directly with the Eleatics. Here, instead, he will draw the consequences of an Eleatic-style argument, if run in a legitimate fashion, taking into account those crucial results about change.

First, after the general framework is set out in terms of the distinction of what holds good of something in its own right, and what holds not in its own right, Aristotle indicates how this distinction can make room for coming to be from what is not.

> We too, on our part, say that nothing comes to be without qualification from what is not. But nevertheless <something> comes to be in some way from what is not – that is, incidentally. (191b13–15)

It is only after this general remark that in the next sentence the slots indicated through this distinction are filled out in terms of the physical principles of the previous chapters. As it turns out, what is not according to itself will be identical to what was earlier introduced as privation:

underlier and shape non-coincidentally, and then again at b25–7, where this claim is repeated about the underlier. (Cf. also the distinction of *Phys.* I.9.192a25–9, where on the first alternative matter is claimed to pass away (and presumably also to come to be) in its own right.)

(For from the privation, which is in itself what is not – an item that is not present as a constituent in <what comes to be>, something comes to be. Yet this is a cause of surprise, and it seems impossible that something should come to be this way, from what is not.) (191b15–17)

Apart from identifying what is not in itself, or what is not according to itself, with privation, these claims also make clear that in some sense there is coming to be from this item. But, as the previous passage insisted, this coming to be cannot be a coming to be without qualification. Or, in the terms in which the distinction was originally introduced above, it cannot be a coming to be from privation in its own right. It can only be a coming to be in a coincidental way from this entity.

This claim then puts the distinction introduced in this chapter squarely in the context of chapter 7. We should recall that there, in the passage stretching from 190b17 to 191a3, the coincidental status of privation was stressed, and contrasted to the status of the underlier, which turned out to be one in number, but dual in form or in kind. This should also mean that there is more to the coincidental status of the role of privation than the fact that privation always resides in an underlier. If the consideration was nothing more than this ontological incompleteness, or non-saturatedness on the part of privation, then the very same consideration would arguably cut both ways. Accordingly, by the lights of the argumentation of chapter 8 we would also be forced to admit that whatever coming to be there is from the underlier, through the unsaturated status of the underlier – from the fact that the underlier always has either privation or form – coming to be from the underlier should also happen in a coincidental manner. Effectively, this would require that the contrast drawn in chapter 7, that coming to be is from the underlier in a non-coincidental way, should be revoked.[11]

One way out of this difficulty would be precisely to return to the considerations of chapter 7 I have just referred to. Those considerations introduced a fundamental asymmetry between the underlier and privation. First, the underlier, as both chapter 7, and now chapter 8, stress, will be present in the result of coming to be, whereas privation has the paradoxical status that it takes part in the process of change only by not being present any longer.[12] Moreover, the underlier, as chapter 7

[11] Moreover, such a consideration in itself would not allow for a difference between the status of privation and form; both of them would come out as incomplete, unsaturated entities.

[12] Ancient commentators, at least from Alexander on, relied heavily on this contrast to block the parallelisms between the status of privation, on the one hand, and of matter and form, on the other.

asserted, is a single entity, with dual characteristics. This is not true of privation. Hence the underlier has more claim on being something from which coming to be proceeds in its own right, even though that entity can also be characterised as in itself an unsaturated one.[13]

IV

Before we can assess these issues in more detail, we need to turn to Aristotle's discussion of the second horn of the Eleatic dilemma, the impossibility of coming to be from what is. In this second part Aristotle suggests that the same distinction, between coming to be from something in its own right and coming to be from something in a coincidental manner, will defuse the Eleatics' argument against coming to be from what is in exactly the same way as he argued for the admissibility of coming to be from what is not in a coincidental manner.

> In the same way <we maintain> that <something> does not come to be from what is, and what is does not come to be <something>, except incidentally. This, too, comes to be this way – the same way as if, e.g., from animal animal might come to be, and from some animal some animal <might come to be>. (191b18–20)

After this the example of such a coincidental coming to be mentioned in the last clause is set out in some detail and some further clarification is added:

> Thus, suppose a dog to come to be from a horse. The dog would not only come to be from some animal, but also from animal, but not in so far as it is an animal, for that is already present. But if something is going to come to be an animal, not incidentally, it will not be from animal: and if <something is going to come to be> what is, <not incidentally, it will> not <be> from what is. Not from what is not either, for we have said what 'from what is not' signifies, namely that <from what is not> in so far as it is what is not. (191b20–6)

We should note three things about Aristotle's argumentative strategy here, all of them distinguishing it from his considerations about the first

See Alexander as quoted by Simplicius, *In Phys.* 238.10–14, and also Alexander, *Quaestio* I.24, p. 38.32–39.1, quoted in n. 15 below.

[13] As I shall argue at the end of this chapter, it is not just that matter is non-saturated. In conformity with the claims about matter, that it can be grasped only through analogy (*Phys.* I.7.191a7–12), it also lacks the distinctness either privation or form confer on it. In other words, it is neither what is according to itself nor what is not according to itself. We should also note that the incomplete and indefinite nature of matter that I am arguing for will loom large in the discussions of ch. 9.

leg of the argument. First of all, this time we get no identification of what is from which coming to be occurs in a coincidental way with an item on the Aristotelian list of principles. Furthermore – unlike on the first leg of the argument – it is not specified whether the being in question from which there is coming to be in a coincidental manner is being according to itself, or not. And finally, the example Aristotle provided arguing for the case of what is coming to be from what is in a coincidental manner, is downright fantastic.

The first two issues are connected. After the first leg of the argument, where Aristotle claimed that from privation, which is what is not according to itself, there is coming to be in a coincidental way, the second leg of the argument could mobilise considerations about the underlier or form, if Aristotle were to identify the being that he speaks about here with one of the principles of his account of change. Both of them, however, have already been designated in *Phys.* I.7.190b17–20 as principles from which the product of coming to be is constituted and comes to be in a non-coincidental manner.[14] Hence, if Aristotle's example provides illumination about cases where some being is the starting point of a coming to be of another being in a coincidental way, this entity cannot possibly be the underlier or the form of the product.[15]

[14] One may try to introduce a distinction here, and claim that the statements of ch. 7 were about the product of coming to be, that that product is constituted and comes to be from the underlier and form in a non-coincidental manner – i.e. in so far as it, the product of generation, is said to be according to what it is (or according to its nature: Aristotle uses *ousia* here at 190b23), whereas the claims of ch. 8 are formulated in different terms, about the metaphysical constituents of the product, that coming to be is from this entity in its own right. This kind of distinction, however, will not work: there is no way to dissociate the product of coming to be and its form, and to claim that the coming to be of the product is non-coincidental according to what the product of the coming to be, having this kind of form, is, but coincidental according to the form conferred on the product.

[15] Note that my argumentation in this paragraph and the following ones runs counter to the exegesis of the majority of the ancient commentators, who try to find a foothold for the second leg of the argument in Aristotle's account of the principles of change. For instance, Alexander in *Quaestio* I.24 at p. 38.14–16 asserts that 'one can divide what is said in the first solution [the one running from 191a34 to b27] so that to say that one part of it is said about privation and the other about matter'. Alexander is able to effect this division by using different criteria for asserting that a coming to be is from an item in its own right (*kath' hauto*) on the one hand, and that it is from an item coincidentally (*kata symbebekos*) on the other. Accordingly, 'the things that come to be come to be from matter in its own right, for it remains the being that it is according to its account and retaining its nature receives the form of what comes into being' (p. 38.32–39.1), but 'generation . . . is not from matter and what is in its own right, if its change to the form is in so far as there is privation' (p. 39.4–7, translations Sharples 1992: 80–2, somewhat adapted). See also Philop. *In Phys.* 171.15–17 and 25–7, and Simpl. *In Phys.* 236.28–32, where Simplicius grafts 'matter in so far as it is being' on the second leg of Aristotle's solution and writes 'what is does not come to be from matter in so far as it is being, but [only] coincidentally, because it is a coincident (*symbebeke*) of

As to the other issue, that these lines do not specify whether the being in question is, or is not what is according to itself, one can venture the following considerations. Form – being the opposite of privation, which is what is not according to itself – would have every reason to be called a being according to itself, or the item that can be designated what is according to itself. But as we have seen, Aristotle's arguments in chapter 7 already ruled out form as a principle of coming to be in a coincidental manner, unless we were to speak about coming to be that is non-coincidental according to a different form – such as a case when a physician turns pale. This is coming to be for the form of physician in a coincidental fashion. In that case, however, not only would the form not be linked to the process in its own right, it could not be characterised as being according to itself in terms of the coming to be we are talking about. In the case of the physician turning pale, the privation from which the coming to be started out was dark. In this case that qualified as what is not according to itself, or the not-being according to itself. Accordingly, in this case the form pale is achieved, hence that form qualifies as what is according to itself, or being according to itself, as far as this coming to be is concerned, not medical art, the form of the physician.

From these considerations we can infer that the being from which Aristotle argues there is coming to be in a coincidental fashion cannot be identical to any of the metaphysical constituents of change, and it cannot be a being according to itself either. This is also borne out by the example with which Aristotle sets out to illuminate his claim. In a case when dog comes from horse, and both of them being animals this item remains, animal is neither underlier, nor privation, nor form, and it definitely does not qualify as that item according to which this coming to be occurs. At the very most it could be identified as the common genus of the two forms – the genus belonging to dog and to horse.[16] But there is no need to go down this road of metaphysical analysis. Instead, it is sufficient to take the Aristotelian example at face value, claiming as it does that when dog

matter that it is not the being that comes to be, as the privation of that being that comes to be is present in it'.

[16] This would situate this item in the Aristotelian metaphysical framework as a common constitutive part of the privation and of the form. The serious drawback of this line of interpretation is that in the parallel case of being coming to be from being in a coincidental manner it would require us to make the highly dubious claim that being too is, if not as a common genus, at least as some common formal component, some constitutive part of the form and the privation.

comes to be from some animal, say horse, it also comes to be from animal – the animal that is this horse.[17]

The advantage of this deflationary reading will be apparent when we turn to the cases of coming to be from being in a coincidental manner, for which Aristotle introduced this example. In every case of coming to be the starting point of this coming to be is being – and the result will also be being. And indeed, just as Aristotle claims about the case in which dog comes to be from horse, the coming to be from what is will not be in its own right as what is.[18] For both of them a starting point would be required that is not the same as the outcome – not-animal for cases of animal generation, like the sperm mentioned earlier in *Phys.* I.7.190b1–5, and what is not instead of those, impossible cases where what is would be generated in a non-coincidental way from something that is. This also accounts for Aristotle's closing remark. After setting out why being cannot be the starting point of a coming to be in its own right, there is reason to return to the other option: if coming to be does not happen from being in its own right, if it happens from something in its own right– as it does from the underlier – we may expect by exclusion that that is not-being. But that conclusion would be unwarranted: coming to be from what is not has already been shown to happen in a coincidental fashion, and the not-being in question is nothing but privation.

Although Aristotle closes this argument here without returning to the status of the underlier, this consideration further reinforces the crucial double-faced characteristic of the underlier, that coming to be happens from it in its own right because it is neither what is not according to itself nor just what is. Accordingly, Aristotle's solution to the Eleatic conundrum about the impossibility of coming to be provides illumination about the status of the underlier, setting his account in very stark contrast to the earlier theories he has been discussing from chapter 4 on. Those earlier accounts took the opposites, and what the opposites qualified, all as beings.

[17] Ross 1936: 496, too, construes the argument in this deflationary way ('"animal" is simply a common predicate of two things with individual differences. And so too with ὄν'), although – as I shall argue presently – this claim about the status of 'animal' is problematic on the emendation he introduces into the text.

[18] This deflationary understanding of the Aristotelian passage may also lie behind Themistius' paraphrase where he writes: 'So too, if water came to be from air, something that is is said to come to be from something that is, not because what already is comes to be, neither because what comes to be already is. Instead from something of which being is predicated something else comes to be to which the self-same being also belongs' (31.4–7, translation from Todd 2012: 48 substantially modified, the most substantial divergence being that I take *auto to on* on line 7 not as 'being itself', but as 'the self-same being', i.e. being of the same type as was present at the start of the coming to be).

Aristotle's discussion of the Eleatic puzzle then makes the crucial contribution that it distances the Aristotelian account from those of these predecessors. It is not only that the opposites they introduce may not qualify for the full status of form, but also that their single underlying substance – like principles of the monists, discussed in *Phys.* I.4.189a33–b3 – does not satisfy the requirements on what it is to be an underlier as implied by Aristotle's refutation of the Eleatics.[19]

At this point I should return to the third issue I have mentioned above, the downright fantastic example Aristotle uses in elucidation of his claim about the coincidental status of coming to be from being. Although the example of dog, an animal, coming to be from horse, another animal, fitted in seamlessly with the argumentative structure I charted above, one nevertheless owes an explanation why a less extraordinary coming to be could not have served as an example. After all, my claim was that practically every instance of coming to be can be described as coming to be from being in a coincidental way. One might have expected that, in that case, Aristotle definitely should not have resorted to the fantastic example he sets out here. Note, however, that although every instance of coming to be can be also described as a coming to be from being in a coincidental fashion, this does not yet provide an example about an item on the analogy of which the claim about coming to be from being in a coincidental fashion could be introduced in an illuminative manner. Most importantly, we need a case where the common element is not the underlier: as I have argued above, the identical factor of the starting point and of the outcome should not bear any metaphysical role in the process of coming to be. Hence the example I mentioned, the physician coming to be pale in a coincidental fashion, may also be misleading, because even though the relevant specification of the underlier for this coming to be is the physician – or the person of whatever expertise – who is dark, the subject term of the sentence picks out the underlier of this case nevertheless. Aristotle's fantastic example is markedly superior, because the starting point and the result are both of them animals, but each of them in its own right, and animal need not refer to anything permanent through the change.[20]

[19] In this regard see Jim Lennox's acute observation in his chapter in this volume (p. 227, with n. 3 there) that Aristotle refers to the principles of his predecessors as matter (*hyle*) and form only in 187a12–23, a context where they are mentioned together with, and contrasted to, Plato.

[20] Note that the claim is not that there need not be anything permanent in the process of coming to be. Rather that Aristotle's example, fantastic though it is, manages to introduce a coming to be where a commonality can be specified without any tacit reference to the underlier.

Nevertheless, dissatisfaction with the example goes back a long way. Simplicius already discusses at *In Phys.* 239.28–240.5 an alternative reading – 'suppose a dog or a horse to come to be' – and he takes it to involve that the dog, or the horse, in question comes to be in the usual way, from dog or from horse. This reading is equivalent to the conjecture introduced by Laas and accepted by Ross in his editions, and by Charlton in the Clarendon Aristotle Series:

> Thus, suppose a dog to come to be <from a dog, or a horse> from a horse. The dog would not only come to be from some animal, but also from animal, but not in so far as it is an animal, for that is already present. (191b20–3)

But resorting to such an emendation and replacing the fantastic example with a regular case of animal reproduction would not meet Aristotle's requirements, as the very claim that animal would be a starting point of the coming to be of the resulting animal in a coincidental fashion would be dubious. This is so, because in the case of dog coming to be from dog and horse coming to be from horse, the generations starting out from the male (and perhaps also from the female) parent of the offspring would certainly qualify as generations in a non-coincidental manner on the species level. These generations occur exactly in so far as these parents are what they are. Consequently, there would be no reason to take these generations to be comings to be in a coincidental fashion even when we refer to these dogs or horses by their genus, that they are animals.[21]

Now that we have set out Aristotle's arguments for the possibility of coming to be from what is and from what is not in a coincidental fashion, we can address the issue, much discussed in recent literature, whether being and not-being here are to be understood in an existential or in a copulative sense, and what difference such a decision might imply.[22]

Although one might be reluctant to make this distinction, as it is admittedly not straightforwardly present in Aristotle's discussion of being, one can definitely make two remarks. One is that in chapters 7 and 8 Aristotle clearly makes a distinction between something being present or absent (*parousia* and *apousia* at 191a7, *enyparchein* at 191b16), irrespective of what claim the item that is present or absent has on being. This is exactly why the role of privation is so paradoxical in coming to be: it is what is not according to itself, even when it is present. The absence of this

[21] See further Waterlow 1982: 17f. with n. 13 there, and Kelsey 2006.
[22] See Code 1976b; Loux 1992; Anagnostopoulos 2013; Clarke 2015.

privation, that is, of what is not according to itself, is 'a cause of surprise, and it seems impossible that something should come to be this way, from what is not' (191b16–17).[23] The same should also trivially apply to form: as the opposite of privation – which is what is not according to itself – it should qualify as what is according to itself, and coming to be can be accounted for by its initial absence, followed by its presence (as Aristotle claims in *Phys.* I.7.191a5–7).

The distinction should apply along exactly the same lines to matter: that constituent is present all through the process of coming to be, but it is a double-faced item, also in the sense that it is neither what is not nor what is according to itself. With all three parallel distinctions in hand one may assert that the being in these characterisations of privation, form and matter, respectively, is not on a par with their presence. In view of this one may venture to claim that presence and absence are very much like our notions of existence and lack thereof, whereas the being involved in the characterisation of privation, form and underlier are nearer to our understanding of copulative being. Before going down this road, however, one must also add that not everything that would qualify as copulative being is admissible in this latter case. Only the privileged type, form, will be admitted as being according to itself. Even though the double-faced nature of the underlier can be truly and explicitly characterised in the copulative sentences that it is 'one in number but not one in form' (190a15–16), or 'one in number but two in form' (190b24), this does not invest it with being according to itself. It remains true that it is neither what is not nor what is according to itself. Moreover, once we take into account that talk about the being of form, as a default, also involves claims about the *presence* of form, we shall see that talk about the being according to itself, as a default, also involves an existential claim.

V

After this we should return to the closing remark about the solution Aristotle supplies in the following lines:

> Furthermore we do not do away with the principle that everything either is or is not. (191b26–7)

[23] Note that the distinction between absence and privation is upheld to the extent that the absence of privation is definitely not the privation of privation.

This is clearly a well-placed jab at those like Plato, who might intend to relegate the domain of coming to be and perishing to a limbo between being and not-being. Such a move would be principally suspect, because then this domain might be characterised in contradictory terms, or in terms which suggest that it is some *sui generis* middle ground between being and not-being.

But we should not only take this assertion as a rallying cry against some Platonists. It should also be checked what its consequences are for the underlier, the true bearer of coming to be in its own right, which is nevertheless neither what is nor what is not according to itself. That would suggest – after acknowledging that the underlier as such has some intrinsic complexity to it – that even though this nature can be grasped only through analogy, it nevertheless is such that any talk about it fully and strictly satisfies the law of excluded middle.[24]

With this Aristotle has completed his discussion of what he put down as 'one way' (191a36) of unpacking the 'only solution' (191a13) to the Eleatic conundrum about coming to be and perishing. This is now immediately followed by 'another way':

> This then is one way [of solving the difficulty]. Another is that the same things can be said according to the potentiality and according to the actuality. This has been explained with greater precision elsewhere. So, just as we said, the difficulties which forced people to do away with some of the things we mentioned are now solved. (191b27–31)

It is not difficult to come up with suggestions how potentiality and actuality could be mobilised against the Eleatic considerations. All it takes is to claim, for the first horn of the dilemma, that an everyday object that is actually not-*F* can also be potentially *F*, and then the Eleatic worry can be stopped: it will be in so far as it is potentially *F* that it can be the starting point of a coming to be *F*. And then the very same case can be mustered against the other horn of the Eleatic dilemma: what is *F* can serve as the starting point for coming to be *F*, provided that the status of the two terminal points of the coming to be are different: the item that is potentially *F* will be suitable as a starting point for a coming to be actually *F*.

This, however, would not do justice to Aristotle's claim that this 'other way' is not a separate answer to the Eleatic conundrum on its own, but rather the 'other way' of unpacking 'the only solution', which uses the

[24] Cf. also Aristotle's charge against the Eleatics in *Phys.* I.2.185b19–25 that their presuppositions will lead to the rejection of the principle of non-contradiction.

results of the preceding discussion of the principles of change. In other words: the deployment of potentiality and actuality is not sufficient in itself for the purpose of defusing the Eleatic puzzle against coming to be. The proper response has to be illuminating about the status of the entities involved in coming to be, and has to give an account of coming to be through an adequate characterisation of these entities.

Therefore one needs to take a further step. If 'the same things can be said according to the potentiality and according to the actuality', one may try to combine the analytic tools of the first half of the chapter, and the tools introduced here, potentiality and actuality. Then the claim would, for example, be that coming to be actually *F* from some actually not-*F* item happens only in a coincidental fashion: this actuality, performing the role of privation, has to reside in some underlier, to the extent that without that it cannot fulfil this function in its own right. Similarly, what is nothing but potentially *F* will not be able to serve, in so far as it is what it is, as the starting point of a coming to be actually *F*. This case, too, is one where the coming to be is in need of another entity: that of the actually not-*F*. It is only the entity that is in some way both of these – very much along the lines of what Aristotle indicated in chapter 7 about the underlier – that can be a starting point of coming to be in so far as it is what it is.[25]

Even with this grafting of considerations about potentiality and actuality on the argumentation of the first part of the chapter all we could achieve was to set out a case where Aristotle in fact *does* use a set of the same fundamental insights in laying the Eleatic challenge to rest. This amounts to less, however, than the claim with which he introduced the arguments of the chapter, that *only* by deploying the results of the analysis of the foregoing chapters *can* the Eleatic challenge be met.

One response to this worry could be to take the claim in a very restricted sense: the way Aristotle addresses Eleatic arguments is the only one to overthrow the Eleatic challenge as contrasted to the ones he mentioned at the end of chapters 2 and 3.[26]

If, on the other hand, one takes the claim in its full, general force, one line of consideration may be the following. In order that there is coming to be and change the entities that serve as the starting points of these changes

[25] See Clarke 2015: 146f.

[26] Plato's response to Eleaticism may have been included among them: see n. 3 above. But even if that is so, ch. 9 will present Plato's account of change on a separate footing, as an attempt – albeit an unsatisfactory one – at grasping the crucially important entity, matter, that is at the heart of the Aristotelian solution. For an assessment of this claim along such considerations see Jim Lennox's contribution to this volume.

upon examination need to turn out to house some complexity. Without that – in a setup in which every entity would have just those features that belong to it in its own right – these entities would invariably manifest themselves, and nothing else. Hence there has to be room for some internal articulation in these entities, so that one can distinguish between the features that are the manifestations of what they are and those other ones that are not the manifestations of the entities themselves.[27]

Granted, this need for some complexity is still not identical to the further, more specific, claim that only the Aristotelian account of principles is able to account for coming to be and change, and hence to give an adequate response to the Eleatic conundrum. But after the piecemeal argumentation of chapters 4–7 Aristotle may have every reason to suppose that his account is the only one that can describe the internal complexity necessary for coming to be and change. Whether he is right in supposing this would require a very thorough examination, which would lead very far afield – perhaps even into a discussion of possible alternatives to the Aristotelian principles themselves.

VI

In closing it makes sense to take stock of what Aristotle seems to have accomplished here. This discussion of Eleaticism provided an opportunity to elaborate further on the status of privation and the underlier. The most important upshot of the chapter in this regard was that although it is the underlier that is the starting point of coming to be in its own right, this status of the underlier is paralleled by the role of privation in coming to be, as a starting point of these same comings to be in a coincidental fashion. Moreover, the characterisation of the roles of these two entities in terms of one of them being a starting point of coming to be in its own right, and the other a starting point of coming to be in a coincidental fashion, was paralleled, as I argued, by the characterisation of their status as what is not: privation was explicitly assigned the status of what is not according to itself, whereas the underlier, consequently, could only qualify as what is not in a coincidental fashion.

This is an important point for appreciating how underlier and privation work in tandem, and how they are related to each other. Furthermore, I have argued that by this characterisation of privation and the underlier Aristotle effectively drives a wedge between his triplet of principles and that

[27] These considerations are analogous to the ones Sarah Broadie discusses in Waterlow 1982: 21–6.

of his predecessors who attributed coming to be to the operation of opposites on a single principle.[28] The Eleatic challenge then turns out to be a crucial test case for assessing the internal complexity of the Aristotelian account, and for showing that in some important sense almost every one of Aristotle's predecessors remained pre-Parmenidean, or at most theoretically on a par with Parmenides. The sole possible exception is Plato, who, albeit insufficiently, made an attempt to formulate an account in a conceptual framework where, at least according to Plato's intentions, the internal complexity of the material principle is to some extent taken on board.

[28] Note that once Aristotle distanced his understanding of the underlier from the 'material' principles of earlier philosophers, he need not keep on stressing this point in other contexts and on other occasions. Most importantly this is the case in *Metaph.* A in the course of Aristotle's rational reconstruction of the development through which his predecessors arrived at his four causes – or at least at some grasp of these causes. There he claims without any caveat that 'most of those who were the first to do philosophy thought that the first principles of all things were solely of the material type' (983b7–8, translation slightly modified from Barney 2012: 76, for further discussion see pp. 76–85 there).

Physics I.9

James G. Lennox

It is generally agreed that Platonism hovers over the final chapter of *Physics* I, although there is little agreement over the exact details of that "hovering" or of Aristotle's attitude toward it.[1] Chapter 9 is not, however, the first time Plato enters the discussion of the principles (ἀρχαί) required for a science (ἐπιστήμη) of nature. In fact, immediately after the lengthy philosophical rebuttal of Eleatic monism in chapters 2 and 3, we read:

> As for the physicists, there are two main ways in which they explain things. Accordingly, making what is a single body, i.e. what underlies – either one of the three, or something else which is more solid than fire but less solid than air – some produce everything else by means of density and rarity, and make [in this way] a plurality of things (these are opposites, and, to put it in general terms, are excess and defect, as are, according to Plato, the great and the small, even though the one [i.e. Plato] makes the great and the small matter, and the one the form, while the others [i.e. the physicists] make the one, i.e. the underlying thing, matter, and make the opposites differences and forms). As for the others, <they argued that> the pairs of opposites are separated out from the one, in which they already were, just like Anaximander says, and also those who say that the things that are are both one and many, just like Empedocles and Anaxagoras. (*Phys*. I.4.187a12–23)

I take this opportunity to thank Diana Quarantotto for organizing the exciting Rome *Physics* I project which eventuated in this volume, and for inviting my participation. This chapter benefited significantly both from the contributions of the other participants in the first *Physics* I workshop in Rome and from their helpful feedback on mine. I would especially like to thank István Bodnár, David Charles, Timothy Clarke, Alan Code, Andrea Falcon, and Diana Quarantotto for both oral and written discussion of this chapter's themes and argument – and in Diana's case, for her skillful editorial suggestions. In addition, I received valuable feedback from Emily Katz and Sean Kelsey, from my students Jon Buttaci, Tom Marré, and Kelson Law, and from colleagues and students in my Fall 2015 seminar on *Physics* I–II at the University of Pittsburgh. Errors that remain despite their best efforts are entirely my responsibility.

[1] Aquinas 1963: 61–6 (Bk. I, Lect. 15, secs. 132–40); Charlton 1992: 81; Giardina 2002: 114–16; Ross 1936: 497, 191b35n; Simpl. *In Phys*. 242.22–243.3; Broadie (forthcoming).

This passage provides a number of interpretative clues which will help us in understanding the argument of this book's last chapter. First of all, it introduces the discussion of the two (different) ways that the natural philosophers (οἱ φυσικοί) speak, treating Plato as one among the group who postulate a single underlying body in addition to opposites, though he apparently treats the opposites he posits, the great and the small, as the underlying matter, rather than as what differentiates a single underlying matter into many distinct things. If that is the view being critically examined in chapter 9, then we should see the beginning of what is, in modern editions, "chapter 9" (but in Aquinas, Lectio 15[2]) as continuing a discussion begun in the immediately preceding paragraphs in "chapter 8."

A second interpretative clue toward understanding chapter 9 provided by this passage in chapter 4 is the fact that it is only when Plato is introduced by name that the language of matter and form is introduced; and only then are the views of the other φυσικοί presented in these terms, by way of *contrasting* their views with those of Plato.[3] This is suggestive of the issues that are foremost in Aristotle's mind in chapter 9; for it looks as if Aristotle sees this Platonic ontology of first principles as the only other serious attempt to distinguish underlying matter from opposites as a means of dealing with the Parmenidean dilemma.

One final interpretative clue from this passage in chapter 4 will aid us in coming to grips with chapter 9: in seeking an abstract characterization (καθόλου, 187a16) of the wide variety of oppositions proposed by his predecessors as *archai*, Aristotle adopts the language of "excess and defect" (ὑπεροχὴ καὶ ἔλλειψις) and then references Plato's opposites, "the great and the small," as an example of the opposition between excess and defect.[4] With the help of a number of discussions in the commentaries on *Physics* I, this points us to a philosophical context in the Academy in which Aristotle is embedded and which, again, helps us understand what underlies the debate being played out in *Phys.* I.9.

Before turning to that chapter, however, it is worth considering another passage, this time in chapter 6, a passage which commentators have correctly

[2] Aquinas divides the modern chs. 2, 3, 4, and 7 into smaller units.

[3] Cf. Kelsey 2010: 109–10. However, Kelsey cites this passage as evidence that Aristotle thinks that the matter/form distinction is not an innovation of his but was common to many of his predecessors (even if the words were not). I would agree about matter, but there are simply too many places where Aristotle treats earlier natural philosophers, other than Plato, as thinking that everything could be accounted for by virtue of material necessity alone to take this passage as Kelsey wants to.

[4] On important mathematical background to these concepts, as reflected in Euclid's *Elements* V, see Sayre 2005: 68–74.

seen as a second reference to Plato's views about what underlies,[5] when the need for a third principle besides the opposites is under consideration:

> But at any rate all of them shape this one [underlying thing] by means of the opposites, by density and rarity, and by the more and less; and these are clearly in general excess and deficiency, just as was said earlier. Also, this opinion, that the one and excess and deficiency are principles of the things that are, seems to be old, though not in the same way, but rather the earliest thinkers say the two act and the one is acted upon, whereas some of the later thinkers say rather the opposite, that the one acts and the two are acted upon. (*Phys.* I.6.189b8–16)

Here certain unnamed later natural philosophers are characterized as treating a pair of opposites, generically characterized as "excess and defect," as the underlying nature which is acted upon by "the one," while earlier natural philosophers held the contrary view that the opposites act on some one thing that underlies. It seems reasonable to include Plato among these later natural philosophers, in light of this being the same view that was explicitly attributed to Plato in chapter 4. But the plural attribution here is significant, and signals that the plural references in chapter 9 should be taken seriously. Moreover, this and a number of related views[6] are attributed, sometimes to Plato alone, but often to him and unnamed others, repeatedly in Aristotle – often enough to justify the assumption that such views were presented and widely discussed within the Academy.[7] Indeed, in a passage we will look at in more detail in discussing chapter 9, Simplicius references a report by Porphyry of a passage in book 11 of Dercyllides' *The Philosophy of Plato* in which he quotes Hermodorus, who attributes to Plato the view that "all those things that are described as great in relation to small possess the more and less" and are therefore "unstable and formless and infinite and what is not."[8] Granted, this is a report at three removes from its source – but the fact that Simplicius provides a very long quotation from it gives us good reason to trust it.

[5] Charlton 1992: 69; Aquinas 1963: 47 (Bk. I, Lect. 11, sec. 94); Broadie (forthcoming): 4–6, 12–15; Simpl. *In Phys.* 204.12–16.

[6] That the material principle is "the indefinite dyad" or "the unlimited," for example.

[7] For example: In *Metaph.* A 6.987b20–1 and 988a11–14 the great and the small are said to constitute a principle in the sense of matter (cf. *Phys.* IV.2.209b35–210a1); and at M 9.1085a7–12 unnamed philosophers are said to make such opposites as "long and short," "broad and narrow," and "deep and shallow" kinds or forms of "great and small" (cf. A 9.992a13–19, N 2.1089b9–14). For further discussion see Katz 2008; Sayre 2005: 66; Ross 1951: 9, 100, 135, 176–82. For a more comprehensive overview of the sort of Platonism that Aristotle has in view in *Phys.* I, see Broadie (forthcoming). The arch-skeptic on this topic is, of course, Cherniss 1944. For his discussion of the relevant passages in *Phys.* I, see 83–123.

[8] Simpl. *In Phys.* 248.5–6, 13–14.

This is a view that Aristotle cites whenever he is drawing attention to the similarities between certain Platonists and certain Pythagoreans.[9] Among these other references, *Phys.* IV.2.209b33–210a2 is perhaps the most interesting discussion for our purposes. What is there presented as Plato's view of place – that it is that which is capable of receiving (presumably forms and numbers) – is first described as the great and the small and then as matter, "just as Plato has written in the *Timaeus*."[10] In fact these are not words that are written in the *Timaeus*. What is one to make of this? One plausible explanation is that Aristotle is referring to *doctrines* that are present in a number of later dialogues, including the *Timaeus*, doctrines which he feels free *to characterize* using a variety of expressions such as "the great and small," "the unlimited," "the more and less," "the indefinite dyad" – all as verbal variants characterizing that which lacks, but is capable of receiving, definition, form, and limit.

This is a plausible explanation in light of passages in Plato's later dialogues, in which we find the same sort of fluidity that we find in Aristotle's characterizations of Plato's views. At *Statesman* 283d1–e1, for example, the Eleatic Stranger presents a division related to the art of measurement that appears to presuppose the ideas to which Aristotle is alluding in the passages we have been discussing:

> [STRANGER] <We should talk> about length and brevity, and excess and deficiency in general. I suppose the art of measurement relates to all of these.
> [YOUNG SOCRATES] Yes.
> [STRANGER] Then let's divide it into two parts; that's what we need toward our present objective.
> [YOUNG SOCRATES] Please tell me how we should divide it.
> [STRANGER] This way: one part will relate to the association of greatness and smallness with each other, the other to what coming into being necessarily is.[11]

Notice the ease with which the Stranger shifts from describing the topic as measurement as it relates to "excess and deficiency in general" to describing one side of the first division in terms of "greatness and

[9] *Metaph.* A 6.987a29–31, b7–14, b20–32.

[10] Ross 1936 (567, ad loc.) is highly critical of Aristotle's treating the receptacle as matter (see too 209b11–13n, 265–6, with the long quote from Taylor's Commentary on the *Timaeus*). If, however, Aristotle is thinking (as is suggested by the question he puts to Plato, why forms and numbers are not in place, 209b34–5) of *Tim.* 53a–b, where the phenomenal traces of the four elements in the receptacle are given "their distinctive shapes, using forms and numbers," it then comes down to how one understands "participation."

[11] Rowe translation, slightly modified. For a detailed discussion of this passage and its context, including an exhaustive review of earlier interpretations and translations, see the appendix to Sayre 2005: 319–51; cf. Gill 2012: 196–7.

smallness" – presumably, as in *Physics* I, this latter opposition is presented as a sort of excess and deficiency, perhaps specified for magnitude.[12] Similarly, in the discussion of the four kinds of beings in the *Philebus* (23c–27c), the initial distinction is into the unlimited and limit, but when it comes to a general characterization of the unlimited, Socrates suggests "the one kind that has the nature of taking on the more and less" (25c10–11);[13] and among those things that "belong to that kind of being which admits of the more and less" we find "greater and smaller" (μεῖζον καὶ σμικρότερον: 25c9–10).[14]

Aristotle, of course, also deploys these oppositional continua for philosophical and scientific purposes; in particular, he typically characterizes perceptible similarities and differences of animals[15] within kinds in terms of "excess and defect" and/or "the more and less," as in this summary remark in *HA* I.1:

> But as a general rule, most of the parts from which the entire bodily bulk is constituted are either the same or differ by the opposites and according to excess and defect; for the more and less one might take as excess and defect. (*HA* I.1.486b13–16)[16]

It is a project for another time to ponder the way in which Aristotle turns these continuous oppositions, which for Plato are the essence of what is limitless and unknowable, into conceptual tools for uniting forms together into a single kind. This much, however, seems clear: Aristotle sides with those of his predecessors who see these opposites as *powers of differentiation and limit*, and against those who see them as representing what is indefinite and unlimited.

The point of mentioning the central place of these concepts within Aristotle's philosophical thought here is simply to note that there is a network of very important metaphysical concepts that are *the shared*

[12] See the references in nn. 7 and 9, above.

[13] Cf. *Phil.* 24e7–25a3, 26d1–3. And cf. the discussion of the role of "the measure" (τὸ μέτρον) in relation to the more and less at *Statesman* 284a8–b2.

[14] I am adopting the translation from Hackforth 1972. The kind of being that admits of the more and less is later termed the Unlimited.

[15] And not just animals: cf. *Metaph.* H 2.1042b21–5, b31–35; *NE* X.3.1173a15–28.

[16] Cf. *PA* I.4.644a17–20: "For as many kinds as differ according to excess and by more and less, these have been brought together into one kind, while as many as are analogous are separated; I mean for example bird differs from bird by the more and according to excess"; and 644b11–15: "For the parts of these [kinds] differ not by analogous likeness … but rather by bodily affections, e.g., greater/smaller, softer/harder, smoother/rougher and such, and generally speaking by the more and less." It is likely no accident that this way of characterizing forms of a kind emerges directly out of a critique of Platonic division and the presentation of continuous, multi-axial division in the preceding two chapters.

currency of those who began their philosophical lives in the Academy during the time when Plato's later dialogues were being composed and discussed. This much, at least, is Aristotle reporting the ideas of late Plato and his associates rather than imposing his own conceptual framework on Plato.[17] That common currency should be seen as part of the background to the discussion of the *archai* of the science of nature in *Physics* I.

In both of these earlier passages in *Physics* I, Aristotle draws our attention to a distinctive feature of Plato's appeal to the great and the small as principles – "the one [i.e. Plato] makes the great and the small matter, and the one the form, while the others [i.e. the physicists] make the one, i.e. the underlying thing, matter, and make the opposites differences (a20) and forms." In this introductory remark in chapter 4 Aristotle is simply laying out the possible positions one can adopt regarding the *archai* of a science of nature and does not take sides on the issue; but in chapter 9 he identifies this very peculiarity of the Platonic approach as a serious mistake, and as an indication that those who adopt this approach have not escaped the Eleatic dilemma, but have simply accepted one horn of it.

I. On to Chapter 9: Some Preliminaries

Before getting down to details about the structure and argument of our chapter, I begin with a suggestion about how it serves both as a fitting conclusion for the discussion to this point and as a proper segue into book II.[18] After the introductory chapter (discussed in the contribution of Andrea Falcon), chapter 2 opens by presenting a range of (possibly exhaustive) positions thinkers might take on how many principles there are and what their character might be (184b15–25). He then takes up the challenge of the Eleatics in chapters 2 and 3 (discussed in the contributions of Timothy Clarke and Laura Castelli), noting that while it is not fitting for one who is in the process of carrying out a scientific investigation of nature to argue with those who deny natural plurality or change, the

[17] Cherniss 1944 takes it that only the *Timaeus* is in the background, and that Aristotle is imposing his metaphysical presuppositions on Plato. He takes no note of the similar language used by Plato in the *Statesman* and the *Philebus*.

[18] There is a well-known controversy, due in part to puzzles about the various titles in the ancient catalogues and in part to the odd repetition of the words that open book II at the end of book I in one of our best mss. (E), as to whether books I and II should be considered as a unit or as quite distinct and unconnected treatises. One of my aims here is to defend a reading of ch. 9 that makes it, as a whole, preparatory for book II. There is an illuminating discussion of this issue in an unpublished ms. by Stephen Menn, '*Physics* II and its relation to *Physics* I–IV' (Menn n.d.).

Eleatic challenge to the whole enterprise is one that must be met;[19] and in truth much of book I is either explicitly (chs. 2, 3, 8) or implicitly (chs. 4–7) directed either to meeting that challenge or to pointing out how attempts to meet it, which lie behind various positions taken by his predecessors, are misguided.[20] It appears to be Aristotle's view that, *with the apparent exception of Plato*, all of his predecessors have been hampered by not explicitly distinguishing between two ways of conceiving a changing thing: [i] as that which underlies changes from one opposite to another; [ii] as one opposite being replaced by another. But by the close of chapter 5 he believes that the discussion has made it apparent that the principles must be opposites (189a10), and chapter 6 is already building the case that if that is true, there needs to be "some third thing which underlies." All of this is his own abstract characterization of what can be learned by studying his predecessors. In chapter 7 he begins to characterize the principles of *opposition* in a new way, as form and privation (190b23–9, 191a8–14); and he begins occasionally to refer to the principle that *underlies* change as matter.[21] As he draws a line on his own account of the *archai* of natural change in chapter 7, he comments in passing that it isn't yet clear "whether the form or that which underlies is substance (οὐσία)" (191a19–20). He is, in other words, shifting the conceptual framework for understanding natural change in the direction of nature as matter and nature as form, i.e. in the direction of the framework presented and defended in book II.

Most of his predecessors, then, attempted to meet the Eleatic challenge, as Aristotle says at the end of chapter 8, without the proper tools. Chapter 9 begins by referring to certain others who *do* make a distinction between matter, form, and opposites, and who might therefore offer a viable alternative to the theory defended in chapter 7. But he goes on to argue that they do so in a confusing and unsatisfactory manner, in part by comparing their views with his own. The chapter divides rather cleanly into [i] discussions of the problems with conceiving the underlying nature, or matter, as Plato and some of his colleagues do, [ii] a discussion of right and wrong ways to think of matter in relation to form,[22] and

[19] To be precise, Parmenides does not speak περὶ φύσεως but *does* state natural *aporiae* so that it is just as well to discuss them, since the investigation is a philosophical one (185a17–20).

[20] See the clear reference to the Parmenidean dilemma at I.4.187a33–b1.

[21] ἡ ὕλη ἀριθμητή, 190b25; ἡ ὕλη, 191a10; and cf. λέγω γὰρ ὕλην τὸ πρῶτον ὑποκείμενον ἑκάστῳ, 192a31. Simplicius also reads ἡ ὑποκειμένη ὕλη at 191a7 (rather than φύσις in our mss.) and omits ἡ ὕλη καὶ at 191a10 in his quotation of the text at 226.6–9. Cf. Diels in Simplicius 1882: 5; Ross 1936: 494.

[22] Cf. the characterization of matter at 192a31–2 with the characterization of the first way in which nature is spoken of at II.1.193a29–30.

[iii] a discussion of the delimited way in which form will be considered by the science of nature (as opposed to first philosophy). The chapter is, then, further preparation for the question of whether matter or form is the most suitable candidate for the nature of natural things – and thus for a radically different investigation of the primary causes and primary principles of natural things, one in which the focus will be much more on inquiry into the causes operative in nature and the proper way to investigate them.[23]

II. Chapter 9: An Interpretative Outline

Aristotle began chapter 8 by saying that he would show that the account of the principles given in chapter 7 was *the only way* to resolve the *aporia* of the earliest thinkers.[24] And that way involved distinguishing between the way in which the matter underlying coming to be *is not* from the way in which the privative opposite *is not* – and of course how each of these is related to the form and that which comes to be. But as we have seen, he has already acknowledged at the beginning of chapter 4 that Plato appears to have similar resources – recall that it is only when referring to Plato's principles that he invokes the contrast between matter and form. Plato too has an underlying substratum, he has the great and the small, and so "opposites," and he makes the One or Unity form. One way to think of this chapter is that it is meeting a challenge, perhaps one he has often heard from colleagues: yours is not the *only* viable response to Parmenides; Plato's later dialogues provide another set of resources.[25]

For purposes of the remainder of the discussion I will divide the text into six unequal segments:

[A] Besides those thinkers discussed in chapter 8, who failed to grasp the underlying nature, Aristotle now introduces others who "touched on it insufficiently" (191b35–6).

[B] He then identifies two errors that lead to their solution being insufficient (191b36–192a2).

[23] In the unpublished paper by Stephen Menn mentioned in n. 18, he worries about the virtually complete neglect of privation in *Physics* II given its importance in *Physics* I – it is mentioned only once, at the end of ch. 1. It is mentioned there during an argument that form is more nature than matter, to which he adds that privation is form in a way. He then puts off for later investigation, perhaps anticipating a possible objection to that claim, the question of whether there is or is not privation in the case of unqualified generation. This, in essence, licenses him to restrict his discussion to nature as matter and as form, especially in a context very much focused on natural substances, their natures, and their causes.

[24] 191a23–4; see Kelsey 2006. [25] Cherniss 1944: 88.

[C] He next compares their three principles with "ours" (192a2–12).
[D] And then compares their approach to matter and form with "ours" (192a13–25).
[E] There is then a discussion of whether what underlies does, or does not, come to be and pass away (192a25–34).
[F] And finally, there is a delimitation of the forms to be investigated by the science of nature as opposed to questions about "the formal principle(s)," which are to be left to first philosophy – and a sentence intended to serve as a bridge to book II (192a34–b4).

[A] 191b35–6

There are certain others who have touched on this, though not sufficiently.

What did these other thinkers touch on – that is, what does "this" (αὐτῆς) refer to in this opening sentence? The best candidate is the nature that is referred to in the previous sentence, at 191b33–4; but then, what nature is *that*? Ross (1936: 497) assumes it is the "underlying nature," which means that the ultimate reference goes back to 191a8. However, Ross, apparently following Simplicius (1882 [Diels]: 242, 10–14), takes a rather catholic reading of "underlying nature," such that it refers to matter, to privation, and to this nature being actually what is not but potentially what is. Here is how Simplicius puts the point, by way of criticizing Alexander's leaner reading:

> But perhaps by 'this very nature' Aristotle does not just mean the nature of privation, but also the nature of matter (and in proceeding Alexander himself points this out) and indeed also the nature of the *per se* and accidental and the nature of the potential and actual about which we are speaking next. (Simpl. *In Phys.* 242.10–14: Mueller trans.)

Without disputing whether these are all ideas that Aristotle could, with suitable qualifications, endorse, I rather doubt that the reference to "nature" at 191b34 can be as ecumenical as Simplicius and Ross suggest. The reference is most plausibly understood as to that which underlies and remains through a change – that is the way "the underlying nature" at 191a8 is to be understood. And while Aristotle sees all of the doctrines listed by Simplicius and Ross as important elements of his response to Parmenides, that doesn't imply that Aristotle's use of "nature" here refers to all of these elements of his account of change.

The next question we need to face about this opening remark is: Who are these "others" whom Aristotle accuses of having insufficiently touched on this nature? Ross (1936: 497) says that the reference at 192a6–8 establishes that it is "Plato and the Platonists" (because of the οἱ δὲ). Simplicius (1882 [Diels]: 242, 23) and Charlton (1992: 81–4), on the other hand, refer only to Plato, and Charlton seems to see the criticisms of this chapter aimed almost exclusively at the *Timaeus*. But Ross is correct that the plural requires an explanation.[26] Given the likelihood that Aristotle and Xenocrates spent time together in Assos, and given the way in which Xenocrates' views are characterized in Theophrastus' *Metaphysics*, he is one likely candidate. But I suspect that when our text was composed Aristotle was part of an ongoing discussion, among people who were, or had recently been, members of the Academy, about ideas that find expression in a number of Plato's later dialogues, in particular the *Philebus*, *Statesman*, and *Sophist*.

> **[B] 191b36–192a2**
> For first, they agree that there is coming to be without qualification from what is not,[27] in which respect <they agree that> Parmenides spoke correctly. Second, it is apparent to them that if <what underlies> is one in number, it is also only one in potentiality – but this is a very different thing.

There is very little manuscript support for the τι in 191b36, and the passage makes tolerably good sense without it. These Platonists agree that there is unqualified coming to be from what is not – but in what respect does this imply that Parmenides spoke correctly? A standard way of understanding this claim is that it reflects a response to the Parmenidean dilemma stated explicitly at the beginning of chapter 8: Coming to be must either be from what is or from what is not. Parmenides famously gave reasons why neither is possible, and thus denied coming to be. These Platonists think that the Parmenidean disjunction that creates the dilemma is correct, but they then accept the second horn of the dilemma.[28]

That interpretation reads a great deal into Aristotle's words, and Simplicius, for one, wasn't buying it.[29] In its support commentators go back to

[26] And see οἱ τὰς ἰδέας λέγοντες (193b36). When Aristotle wants to restrict the ideas he is criticizing to Plato, or to the *Timaeus* in particular, he tends to be careful to do so. For example, later in the *Physics* at 203a4–16 views he ascribes to "the Pythagoreans" are consistently compared with views of Plato. Similarly at 209b11–13 and 209b33–210a2 he reports what Plato says in the *Timaeus*, while at 209b14–16 he reports what Plato opines ἐν τοῖς λεγομένοις ἀγράφοις.

[27] Omitting the τι, which is absent from the best mss.

[28] Ross 1936: 497, 192a1n; Charlton 1992: 81. Mueller actually built this interpretation into his translation of Simplicius' quotation of 191b35: "... and that that [part] was correct in Parmenides' formulation [of his dilemma]" (Baltussen, Share, Atkinson and Mueller 2012: 126).

[29] More recently, it is also questioned by Broadie (forthcoming: 2), though on different grounds that those defended here.

the beginning of chapter 8. But in our passage what Parmenides is said to have stated correctly is that there is coming to be without qualification from what is not. In what sense can that view be attributed to Parmenides?

Looking to the opening of chapter 8, we see that there is an asymmetry to the way in which the two horns of the Parmenidean dilemma are expressed. The statement of the first horn is straightforward: what is cannot come to be, since it already is; the second horn, on the other hand, is expressed in an unusual way: nothing can come to be out of what is not, *since there must be something underlying* (191a25–31). That is, in order for coming to be to occur, there must be something underlying to come to be. But since the Platonists under consideration in chapter 9 are supposed to be insisting that that which underlies *is* what is not, something *can* come to be out of what is not. The Eleatic position as stated at the outset of chapter 8, these Platonists would insist, was correct – the Eleatics just need to go on to recognize that what underlies is what is not. This appears to be what Simplicius has in mind:

> However, Plato, considering what is not in the sense of what is other <as in the *Sophist*>, accepted [Parmenides'] minor premise <what is anything else besides being is not>, but rejected the major, the one which says that what is not is nothing. (Mueller trans. 2012: 128)

Simplicius' thought seems to be this: the Parmenidean argument, which concludes that there is nothing other than what is, has two premises, a major and a minor. The major premise is "What is not is nothing at all"; the minor premise is "What is other than what is is what is not." The Plato of the *Sophist* can accept this minor premise, because he has made a place in his ontology for what is not in the sense of what is other than what is. Though for reasons having to do with his desire to reconcile differences between Aristotle and Plato, Simplicius ends his comment on this lemma by expressing dissatisfaction with this reading, it is preferable to building in a reference to a dilemma that is simply not there.

We can now move on to the final puzzle in this passage: the claim that the Platonists think it follows that "if <what underlies> is one in number, it is also only one in potentiality."[30] Commentators have struggled with how to understand the conflation that Aristotle is attributing to these Platonists.

[30] It is unclear what the subject of ἐστὶν at 192a2 is: Ross 1936: 347 takes it to be "the substratum"; Charlton 1992: 82 and Aquinas 1963: Lect. 15, sec. 130 treat it as completely general – these Platonists think that if any two things are the same in number then they are the same "in possibility." Pellegrin (2002) takes the subject to be "cette nature," thus (I take it), in effect agreeing with Ross.

Ross (1936: 497) interprets δύναμει μία as "one in significance" – and then, to further confuse matters, writes: "δύναμει = the *eidei* of 190a16, b24." Why, then, would Aristotle not have just written the same thing he had written on those previous occasions; and in what sense are being one in significance, one in form, and one in potency equivalent expressions?

At 190a16 and b24, Aristotle is drawing a distinction between what is one in number yet distinct in form or *logos*. And to my knowledge Aristotle never equates *that* distinction with a distinction between what is one in number and yet distinct in potentiality. Ross is not alone in making this sort of move: Aquinas, though eventually providing a more reasonable interpretation of the passage, translates δύναμις here as *ratio*, which is his usual translation for λόγος;[31] Charlton *translates* "possibility" but in his commentary (1992: 82) follows Ross and Aquinas, saying "i.e. in account." In short, the commentaries on this passage present us with a number of confusions.

In the context of his disagreement with the Platonists, Aristotle appears to be suggesting that they fail to distinguish, within numerically the same subject, that subject *qua lacking* (but *being capable* of possessing) a certain form and *qua* the *material* for that form. But (to use Aquinas' example) lacking shape and being bronze, while both ways of potentially being a statue, are not so in the same respect. Thus, while it is correct to say there are two distinct accounts of that which is coming to be, the error that Aristotle is *specifically* concerned with is that which identifies the *indeterminate potency* of the subject *qua* privation with the much more *determinate and goal-oriented potency* of the subject *qua* matter. That is, Aristotle is already looking toward one central idea of book II – that the material of a natural object *is* its nature, though not in the same respect as its form, for form can be identified with the goal of coming to be.

[C] 192a2–12

For we ourselves say that matter and privation are different, and that of these one, the matter, is what is not incidentally, while privation is in itself what is not; and that the one, the matter, is near substantial being and is substantial being in a way, while the other is in no way at all substantial being; but these other philosophers <say that> the great and the small are alike what is not, either the two taken together or each one separately.[32] So this type of triad and the previous one <of ours> are completely different. They advanced to the point <of recognizing> that there needs to be a certain nature to underlie, yet they make this one. For even if

31 To compound the confusion, the English translators render Aquinas' *ratio* here as "nature"!

32 τὸ μὴ ὂν could be taken as grammatical subject. I am taking "the great and small" as the subject because the issue at stake is what should be called "what is not" and in what way.

someone makes it a dyad, calling it great and small, nonetheless it comes to the same, for he overlooks the other nature.[33]

Aristotle acknowledges that both he and these Platonists[34] have triadic ontologies that superficially look similar, and thus ought to be able to deal with the Eleatics in a similar manner: they both appear to have something that slots into the underlying matter/privation/form framework. But the appearance is deceptive: we know from chapter 4 that Plato in particular (and he apparently had followers in this) somehow *identifies* the opposition of the great and the small with the underlying material, which according to the apparent reference to the same doctrine in chapter 6 he conceives of as passive and somehow as acted upon by Unity.[35] Moreover, if they do this, then the appearance that "the great and the small" are a *pair of opposites* in the same sense that they are in certain pre-Socratics is illusory. This expression rather seems to refer to a single *indefinite and unbounded continuum*, awaiting definition and determination by the application of a unit of measurement, such that one could say both what a differentiated part was in itself and what it was equal to. This might, in turn, be what these thinkers have in mind in identifying Unity with Form.

These are, of course, ideas that Aristotle refers to a number of times elsewhere. In *Metaphysics* A, in the context of insisting that Plato only makes use of two of the causes discussed in the *Physics*, Aristotle says: "And it is apparent that the underlying matter, of which the forms are predicated in the case of the perceptible things and the One in the case of the forms, is a dyad, the great and the small" (*Metaph.* A 6.988a11–14).[36] Aristotle, however, and whomever it is that "we" refers to in this chapter,[37] distinguishes sharply between the two kinds of potency, exemplified in matter

[33] I am supposing that τὴν ἑτέραν at a12 is a reference to the overlooked στέρησις (cf. Ross 1936: 497). On στέρησις as nature in a way, see 193b18–20. Aristotle has just attributed to the Platonists the view that they see the need for an underlying nature, but do not differentiate matter and privation in what underlies.

[34] Plato must be included, since, as we have seen, this very doctrine is attributed to Plato by name a number of times, both in the *Physics* and *Metaphysics*. And in both texts Aristotle stresses that in this respect Plato's ontology and that of the Pythagoreans are very similar to each other.

[35] Cf. *Phil.* 26e6–8, where it is suggested that τὸ ποιοῦν καὶ τὸ αἴτιον would rightly be termed Unity (ἕν).

[36] Cf. *Metaph.* A 6.987b20–2, 7.988b5. It is beyond the scope of this chapter to ponder the idea that in the case of Forms the One is predicated of the great and the small as their underlying nature. Cf. Sayre 2005: 92–8.

[37] The first person plural is also used at 189b30 as he begins to present his alternative account of the principles of natural inquiry. At 192a3, a6 what "we ourselves say" is contrasted with what "they" (the Platonists) say. This is in marked contrast to *Metaph.* A 9 (990b9, b17, b24, 992a11) where, in the midst of a scathing critique of the theory of forms, he uses the first person plural to identify those who posit forms!

and in privation. That which underlies, understood as matter, is "near to substantial being (οὐσία) and is substantial being (οὐσία) in a way." In *Metaph.* Η 1–2 Aristotle repeatedly acknowledges that what underlies is οὐσία in a way: "It is evident that matter too is substantial being, since in all changes from one opposite to the other there is something which underlies the change."[38] On the other hand, a privation, just in so far as it is a privation, is simply a lack, an absence. In identifying the great and the small, or the more and less, or excess and deficiency, with what underlies, and with what is not, Aristotle takes it that the Platonists do not, in fact, have the requisite distinctions between that which underlies, privation, and form to deal with Parmenides.

Commenting on this stretch of text, Simplicius tells us that Porphyry reported on a quote from Hermodorus in Dercyllides' *The Philosophy of Plato.* Fortunately (since the original is lost) Simplicius quotes the passage. The bit on which I want to focus our attention goes as follows:

> And all those [things] which are described as great in relation to small possess the more and less. For it is by the more that they are greater and smaller, going on without limit, and likewise too broader and narrower, heavier and lighter, and everything which is expressed this way will go on without limit. (Simpl. *In Phys.* 248.6–9)

The quoted passage goes on to identify "the more and less" first with what is "unstable and formless and infinite and what is not" and then, possibly, with matter.[39] This passage is of significance for understanding the context of the debate being played out in this final section of *Physics* I. Hermodorus met Plato in Syracuse and studied in the Academy until Plato's death, after which he wrote a biography of Plato.[40] As John Dillon points out, in this passage Hermodorus appears to be comfortable ascribing the identification of the great and the small (and the more and less) with matter to Plato. To quote Dillon:

> We thus have here, attested by Hermodorus, independently of Aristotle, in the first generation after Plato's death, confirmation of the two supreme principles of Plato's so-called "unwritten doctrines" (though indeed they are fairly plainly written about in the *Philebus*), and the unhesitating identification of the "unlimited," passive principle as "matter." (Dillon 2003: 202)

[38] 1042a32–4; cf. 1042a26–8, 1042b9–11.

[39] The last sentence quoted by Simplicius is this: "But matter is not a principle, and therefore also it was said by those around Plato that there is one principle" (248.17–18). Simplicius then goes on to say that "it has . . . become clear from these words in what way Plato called matter great and small and what is not" (248.19–21).

[40] Dillon 2003: 189–204.

As we saw earlier, Aristotle, in seeking to find an abstract way of characterizing the various oppositions that he finds in his predecessors as principles of nature, repeatedly uses the language of "more and less" and "excess and defect" – stressing that in most cases his predecessors take their opposites to be principles of differentiation,[41] while the Platonists identify *their* opposites, the great and the small, with the unlimited and the indefinite. As the passage quoted in note 41 from *Metaph.* H 2 makes clear, on this point Aristotle stands firmly with those who treat opposites as "principles of being" and thus in fundamental opposition to the Platonists he is here discussing.

As I also noted earlier, in *Phys.* IV.2.209b35–210a2, Aristotle goes so far as to ascribe to the *Timaeus* the view that place, as that which is capable of receiving forms and numbers, is both the great and the small and matter.[42] I argued there that Aristotle likely is referring to doctrines present in a number of later dialogues, including the *Timaeus*, and feels free, as Plato himself did, to refer to those doctrines with a number of different expressions. The quotation from Hermodorus is some evidence that this was a common practice among Academics of the time.

> **[D] 192a13–25**
> The nature which remains is a joint cause with the shape of things that come to be, as a mother; the other part of the opposition might often be imagined not to exist at all, if one directed one's thought to its maleficence. For, supposing there is a certain thing that is divine, good and desirable,[43] we say there to be one thing that is its opposite, and another which by nature desires it and reaches out for it in accordance with its nature. For them, on the other hand, it results that the opposite reaches out for its own ceasing to be. And yet neither is the form able to desire itself, on account of not being in need, nor is the opposite (for opposites destroy each other), but this is the matter, just as if the female reached out for the male and the ugly for the beautiful – except that in itself it is neither ugly nor female, but only incidentally so.

As has often been pointed out, the reference to a joint cause (συναιτία) and to a mother (μήτηρ), in the context of a discussion of the material and formal principles of the Platonists, suggests that the *Timaeus* is in the

[41] A view with which Aristotle agrees; cf. *Metaph.* H 2.1042b21–4. At 1042b31–4 he is emphatic: "We should, then, grasp the kinds of differentiae, as these differentiae will be principles of being. For instance, there are things differentiated by more and less, or by dense and rare, or others of this sort, and all these are cases of excess and defect."

[42] See p. 227, above.

[43] It is not clear how to take the genitive absolute here. The tone of the passage is quite different if one reads it as concessive: "For supposing that . . . ."

background.[44] On the other hand, Aristotle here seems clearly to be presenting his own approach to understanding matter, form, and privation in the context of coming to be: Why would he express his own position in a way that brings Plato's view to mind? What I have been suggesting throughout this discussion is that Aristotle is here in an ongoing dialogue with certain Platonists who know his views on the topic of how to respond to the Eleatics and feel that Plato has a viable alternative. In this chapter he is expressing his views about the principles of natural change in a way that will be clear *to them*, and will bring his (or "our") views into contact with theirs. Here, for the first time explicitly, the *teleological* relationship between matter and form is being stressed, but described in terms of desiring and reaching out for what is divine, good, and desirable; that which remains through the change is a joint cause (συναιτία), like a mother, along with the form of what is coming to be. By doing so he is able to stress the importance of distinguishing between what underlies *qua* potential for form and *qua* privation. If you don't make that distinction, then what underlies ends up reaching out for its own destruction.

Moreover, it is not as if the metaphor being appealed to here is un-Aristotelian. In his account of animal generation, Aristotle treats the female's contribution to generation as the matter and as passive, the male's contribution as form and active, allowing the generator to participate in the eternal and divine.[45] This way of characterizing the relationship between matter and form during change is, then, likely to resonate with all parties to the debate. Even the criticisms of equating that which underlies with the privation are expressed in ways that will make Platonists see what is at stake if one makes this error. Read in this way, the very last comment is both true and mildly humorous: of course Aristotle sees the female principle in generation as "material" and the male principle as providing the form, and what lacks form as "ugly" by comparison with what has it. But characterized at a higher level of abstraction, the underlying principle of change is neither female nor ugly, though in a particular case of change it may be appropriately so described.

As I read this section of the text then, the argument runs along these lines: The Platonists did do one thing right; they recognized the need, in dealing with the Eleatic dilemma, of a nature that underlies and remains through change. But they failed to distinguish the role of opposition from

44 For other possible verbal and philosophical echoes see Charlton 1992: 82–5; Cherniss 1944 takes it to be virtually the only text on Aristotle's mind.

45 *GA* I.21.729b12–16, 730a24–8, II.4.738b20–7; *DA* II.4.415a26–b8.

the role of the underlying nature in accounting for change. They may posit an opposition, like the great and the small, excess and defect, or the more and less: but they equate this opposition, in terms of its role, with the underlying nature. So (in the terms of the *Timaeus* or *Philebus*) the underlying nature and the forms are joint causes[46] of coming to be, and approaching things as they do, you might treat the other part of the opposition simply as what is not. But if you then equate the continuum of opposition with that which underlies, all sorts of unfortunate consequences arise: the material nature will strive for its own destruction, since once the form comes to be the privation is destroyed. Putting the matter in somewhat Platonic terms, here is how *we* look at it. It is the underlying material that naturally reaches out for the form, i.e. for that which is divine, good, and beautiful. Certainly in the process the privation is transformed into its opposite, the form; but the enduring material nature is simply realizing its potential, in a way realizing its nature.

> **[E] 192a25–34**
> In a way it is possible that what underlies ceases to be and comes to be, and in a way not. Considered as that in which, it ceases to be in its own right (for that which ceases to be, the privation, is in this); but considered as potential, it does not cease to be in its own right, but it is necessary that it be incapable of ceasing to be or coming to be. For if it came to be, there would need to be something prior to underlie from which, as a constituent, <it came to be>; but this is its nature[47], so it will be before it comes to be (for by matter I mean the primary underlying subject for each thing, from which, as a constituent, something comes to be non-incidentally). And if it ceased to be it would finally come to this, so it will have ceased to be before it had ceased to be.

Aristotle is now prepared to return to the model of change established in I.7; you can think of that which is coming to be from the perspective of being that in which the privation is present, the uncultured human being or the amorphous lump of clay. During the process of a child becoming educated or a lump of clay becoming a bowl, the uncultured or amorphous thing passes away. Or you can think of that which is coming to be from the perspective of its potential for becoming informed and thereby as the underlying subject of unqualified coming to be. During the process of a child becoming an educated person or of a lump of clay becoming a bowl, each realizes their potential, but the underlying subject remains

[46] Which characterizes necessity in the *Timaeus*, and is so perfectly fine as a way to characterize the material nature here.

[47] Reading αὐτῆς with IP.

(cf. I.7.190a17–26); the bowl and the educated person come to be, but the clay and the person do not come to be.[48]

A question about this passage that has bothered commentators from Simplicius (252.26–8) to Charlton (who actually proposes a textual emendation to deal with it (1992: 83)) concerns the meaning of the phrase "that in which" (τὸ ἐν ᾧ). When Aristotle claims that "considered as that in which (τὸ ἐν ᾧ), it [that which underlies] ceases to be in its own right (for that which ceases to be, the privation, is in this)," in what sense does the underlying thing cease to be in its own right? Simplicius claims that Aristotle "misapplies 'that in which' to what is in the substratum," and Charlton notes that this "is an odd phrase when what is supposed to be 'in' the thing is a privation . . . and as 'that in which the privation is' [as Ross 1936: 498 and Aquinas 1963: Lectio 15, sec. 139 take it], the matter would surely pass away rather by virtue of concurrence than of itself (a26)." Charlton's solution is desperate, but suggestive: desperate because without any justification in the mss. or commentators he wishes to delete the entire parenthetical explication as a confused gloss; but suggestive in that he points out that the expression τὸ ἐν ᾧ γίγνεται is used in the *Timaeus* (50d1) to characterize the receptacle, just prior to the metaphorical comparison of it to the mother, a metaphor Aristotle alluded to at 192a15. Charlton takes this to license reading τὸ ἐν ᾧ as a reference to the privative character, since he reads the whole of 192a6–25 as full of ironic references to the *Timaeus*. However, given that Aristotle appears to be adopting modes of expression that are evocative of the Academy, it is reasonable to suppose that this expression refers to the "receptacle" characterized *solely* in terms of what is "in" it. How else might we characterize it, other than by some sort of "bastard reasoning"? Here again, then, Aristotle is making his point in a way that he hopes will strike home with those who have found the Platonic alternative attractive. When that which underlies is characterized solely by reference to the privation present before the change, it ceases to be (under that description) in its own right. But characterized as the primary underlying subject present in that which comes to be, i.e. as matter, it neither comes to be nor passes away.

> **[F] 192a34–b4**
> Concerning the principle according to the form, whether it is one or <they are> many and what it is or what they are, it is the work of first philosophy to determine with precision – so let us put <such questions> aside until that occasion. In the following discussions we will be speaking about the

48 Of course there is a well-known problem of how to deal with biological generation on this model.

> forms of natural and perishable things.[49] So, then, let us take it as determined in this manner that there are principles and which and how many in number they are; but let us carry on the discussion, beginning from another starting point.

Aristotle concludes *Phys.* II.2 by saying that it is the work of first philosophy to determine the status of "the separable and what it is" (194b14–15).[50] That comment follows him saying that it is the job of the natural scientist to know "the form" and "the what it is" *up to a point* – specifically, up to the point of knowing it as that which a particular sort of matter is for the sake of, and knowing it as "separable in form, but in matter" (194b9–13). I see this concluding paragraph of book I as preparing the ground for this limited investigation of form by the natural scientist in book II. Both here and in II.2, then, Aristotle has the thought that there are different sorts of questions about the nature and status of form that are appropriate for first philosophy and for natural science. I have nothing interesting to say about what it might mean to explore questions about whether the formal principle is one or many, although the *Metaphysics* has a good deal to say about what the formal principle is according to Aristotle and how it differs from Plato's understanding of form.

In any case, we are here concerned with the first book of the *Physics*, and here, how it might relate to the rest of that work. Here is a list of questions that ought to be answered by someone about to engage in *natural* inquiry, none of which have been answered in book I:

1. What is a nature?
2. What is a natural being?
3. Is it the underlying matter or the form that constitutes the nature of a natural thing – or is it somehow both?
4. Does one of these candidates for nature have priority over the other?
5. Are these *principles* of change (if that is what they are) also *causes* and *elements* of natural change (thinking back to 184a10–14)?
6. Are there other causes of natural change?

All of these questions, as it turns out, are (or are related to) questions about form; and judging by the first three chapters of book II, they are appropriate questions to be explored by the natural scientist. But as of the

[49] I defend the text and translation here in Lennox 2008: 153 n. 4, and discuss the options in the appendix, below.

[50] Which can, of course, be taken in a number of ways. Some read it as referring to a discussion of what "the separable is"; it may also be referring to a discussion of both separability and essence. This is not something that needs to be decided here.

end of book I, while the term *matter* is applied to the underlying nature, the idea that form too might be a nature has yet to be broached, let alone defended. But the way has been prepared for one key ingredient in the defense of the idea that nature refers to both the form and the matter of natural things (194a12–15, 199a30–2) – namely, the relation of form and goal (194a26–30, 198a24–7, 199a30–2). When we are told that "that which by nature desires and reaches out for [the divine, good and desirable] in accordance with its nature" is matter, and that "form doesn't reach out for itself," the idea of the teleological unity of matter and form, so elegantly expressed in the following passage, is prefigured:

> And since nature is twofold, the one as matter and the other as shape, and this is an end, while the other things are for the sake of the end, this [nature] would be the cause in the sense of that for the sake of which. (*Phys.* II.8.199a30–2)

Appendix on 192b1–2

At 192b1–2, the Λ group of mss., reflected in Bekker's printed text, has περὶ τῶν φυσικῶν καὶ τῶν φθαρτῶν εἰδῶν. Ross, following ms. E, omits the second τῶν and thus suggests taking the first with εἰδῶν. Since Aristotle argues in the *Metaphysics* that it is the generated composite and not the natural forms that comes to be (*Metaph.* Z 8.1033b5–9, b16-21, Z 9.1034b7–10, H 5.1044b21–4) and the same arguments that lead to that conclusion should lead to the conclusion that forms are imperishable, a reading that avoids attributing to Aristotle a belief in perishable forms is desirable, and there are two ways of doing so. One is to read the text as printed by Bekker and translate "concerning forms of natural and perishable things," taking the genitive plural of εἶδος to be governed by περί and taking φυσικῶν and φθαρτῶν substantivally. If one follows Ross, it is still reasonable, given Aristotle's Greek, to treat them as substantives, even in the absence of definite articles. Moreover, since Aristotle believes that there are both perishable and eternal natural things, the καὶ should not be rendered as "i.e.," as it is in the Hardie and Gaye translation.[51] The addition of "perishable" here may be intended to delimit the discussion to forms of those natural substances listed at the beginning of book II: animals, their parts, plants, and the four elements.

[51] Hardie and Gaye 1930.

Physics Book I

Translation

This translation is a joint work: each contributor translated her or his chapter of Physics *I; Lindsay Judson and Diana Quarantotto coordinated the entire translation with the aim of making it as uniform as possible, at least as far as the most important terms and expressions are concerned. The translation of* τὸ ὄν *is a partial exception. This expression is generally translated as 'what is', but in certain places in chapters 2 and 3 it is also rendered as 'being'. Timothy Clarke and Laura M. Castelli use these two different translations to reflect the fact that, in these chapters, the expression sometimes refers to what has being ('what is'), and sometimes to the property or predicate itself ('being'). One occurrence of* τὸ ὄν *in chapter 6 is translated as 'that which is': see Alan Code's footnote to 189a13.* Οὐσία *is generally translated as 'substance', but as 'nature' in chapter 7 at 190b19 and as 'substantial being' in chapter 9. The translation of* ἐναντίον, ἐναντία *and* ἐναντίωσις *as 'opposite', 'opposites' and 'opposition' is argued for by Lindsay Judson in his chapter; all the contributors agree with his reasons and have adopted this translation in their chapters.*

1. (**184a10**) Since knowledge and science come about, in all the investigations in which there are principles or causes or elements, from grasping these things (for it is then that we think we know each thing, whenever we have grasped the first causes and the first principles and <got> as far as the elements), it is clear that in (**a15**) the science of nature too one should try to determine, first of all, the things concerning the principles. The natural road is from what is more knowable and clearer to us to what is clearer and more knowable by nature: for it is not the case that the same things are knowable[1]

[1] *OCT*: γνώριμα. E: γνωριμώτερα.

to us and also <knowable> without qualification. For this very reason, it is necessary to proceed in this way from what is unclearer (**a20**) by nature but clearer to us to what is clearer by nature and more knowable. What is initially evident and clear to us are the things that are confounded to a degree: it is only later, starting from these [confounded] things, that the elements and the principles come to be known to those who divide them. For this reason, one must advance from the universals to[2] the particulars: for the whole (**a25**) is more knowable in relation to perception, and the universal is a sort of whole, for the universal comprises many things as[3] its parts. (**184b1**) The same thing happens in a way to names in relation to their definition: <a name> signifies some whole in an indefinite way, e.g. 'circle', whereas its definition divides it into the particulars. Children too at first call[4] all men fathers and all women mothers, and later distinguish each of the two.

2. (**b15**) There must either be one principle or more than one. If there is one, it must either be unchanging, as Parmenides and Melissus say, or changing, as the natural philosophers say, some saying that the first principle is air, others that it is water. And if there is more than one, there must either be a limited or an unlimited number. If a limited number greater than one, there must either be two or three or four (**b20**) or some other number. And if there are an unlimited number, then either, as Democritus says, the genus is one and they are distinguished by shape or by species, or they are also opposites.[5] And those who inquire into the number of the things that are are also inquiring in a similar way. For they inquire primarily[6] into what the things that are are from, asking whether these things are one or many, and if they are many, whether there are a limited or unlimited number, so that (**b25**) they are inquiring into whether the principles and the elements are one or many.

Now, to investigate whether what is is one and unchanging is not to investigate into nature. (**185a1**) For, just as for the geometer too there is no longer any argument to give against an opponent who destroys the principles, but this is instead something either for another science or for one common to all, so too for the person investigating principles. For there is no longer any principle if it is only one, and one in this way; for a principle is a principle *of* some thing or things. (**a5**) So, to investigate whether it is one in this way is like arguing dialectically against any other thesis put forward for the sake of argument (like the Heraclitean thesis, or if someone should

[2] *OCT*: ἐπὶ. E: εἰς. [3] *OCT*: ὡς. E: ὥσπερ. [4] *OCT*: προσαγορεύει. E, V, I: ὑπολαμβάνει.

[5] Reading σχήματι δὲ ἢ εἴδει διαφερούσας with E, I, Simplicius and Themistius.

[6] Reading ζητοῦσι πρῶτον with F, I and J.

say that what is is one human being), or like solving an eristic argument, which is just what both arguments contain, both Melissus' and Parmenides'. For they assume falsehoods, (**a10**) and are not deductive. Or rather, the argument of Melissus is crude and contains no difficulty – grant him one absurdity and the others follow: this is not very hard. But, for our part, let it be assumed that natural things, either all or some of them, undergo change. This is clear from induction. And at the same time nor does it belong <to us> to solve everything; (**a15**) only those which someone falsely proves from the principles, but not others, just as it is the task of the geometer to solve the quadrature by way of segments, but not the quadrature of Antiphon. However, although <the Eleatics do> not <speak> about nature, they nonetheless happen to state physical difficulties.[7] So it is presumably a good idea to have a little dialectical discussion (**a20**) about them. For the investigation does contain some philosophy.

Since being is said in many ways, the most appropriate starting-point of all is to ask in what way those who say that 'all things are one' speak <of being> – whether all things are substance, or quantities, or qualities, and again whether all things are one substance, like one human being, or one horse, or one soul, (**a25**) or whether all things are quality, and this is one, like pale or hot or one of the other things of this sort. For all these differ a great deal, and all are impossible to maintain. For if, on the one hand, there is substance and quality and quantity, then whether these things are detached from one another or not, the things that are will be many. But if, on the other hand, all things are quality or quantity, then whether substance is (**a30**) or is not, this is absurd, if one should call the impossible absurd. For none of the others is separate apart from substance. For all <the others> are said of substance as an underlying thing. And Melissus says that what is is unlimited. Therefore what is is a quantity. For the unlimited is in the <category of> quantity, and it is not possible for a substance or a quality or an affection to be unlimited, (**185b1**) except incidentally, if they are at the same time also certain quantities. For the account of the unlimited employs quantity, but not substance or quality. If, therefore, it is both a substance and a quantity, what is is two and not one. But if it is substance alone, (**b5**) then it is not unlimited, nor will it even have any magnitude at all. For then it will be a quantity.

Further, since the one itself is also said in many ways, just as being is, it is necessary to investigate in what way they say that the universe is one. And

[7] Following Ross's punctuation: οὐ μὴν ἀλλ' ἐπειδὴ περὶ φύσεως μὲν οὔ, φυσικὰς δὲ ἀπορίας συμβαίνει λέγειν αὐτοῖς.

we call one either the continuous, or the indivisible, or those things of which the account of their essence is one and the same, such as μέθυ and οἶνος.[8] Now, if it is (**b10**) continuous, the one is many. For the continuous is divisible to infinity. (And there is a difficulty concerning the part and the whole, although presumably it is not <a difficulty> for the argument, but <a difficulty> in its own right. That is, whether the part and the whole are one or more than one, and how they are one or more than one, and if they are more, how they are more. This also applies with regard to non-continuous parts. (**b15**) And if each part is one with the whole by being indivisible <with respect to the whole>, then <there is the difficulty> that they [*sc.* the parts] also bear this relation to one another.) But if it is one by being indivisible, then nothing will be a quantity or a quality, nor then will what is be unlimited, as Melissus says, nor limited, as Parmenides says. For the limit is indivisible, but not the limited thing. But if everything that is is one in account, (**b20**) as are mantle and cloak, then it follows that they are affirming the account of Heraclitus. For being good and being bad will be the same, and being good and being not-good, so that the same thing will be good and not-good, and a human and a horse, and their account will not be about the fact that the things that are are one, but about the fact that they are (**b25**) nothing.[9] And being this quality and being this quantity will be the same. And the more recent of the early thinkers were also troubled lest the same thing should turn out for them to be at the same time both one and many. This is why some people took away the 'is', like Lycophron, while others refashioned their speech, saying not that the human 'is pale', but that he 'has-paled', and not that he 'is walking', (**b30**) but that he 'walks', so that they would never make the one be many by adding the 'is', supposing that the one or being is said in only one way. But the things that are are many either in account (for example, being pale and being educated are different, yet the same thing is both; therefore the one is many), or by division (as with the whole and the parts). But here (**186a1**) they were already in difficulty, and they conceded that the one is many, as though it were not possible for the same thing to be both one and many – although not those that are opposed. For the one is both potentially and actually.

3. So, if one proceeds in this way, it becomes apparent that it is impossible that (**a5**) things that are be one, and it is not difficult to solve the arguments from which they argue for their claims. For both of them,

[8] Two Greek words for wine.

[9] Removing Ross's dashes separating off ὥστε . . . μηδέν; adding a comma after the second εἶναι at 185b22; and adding a full stop after μηδέν.

i.e. Melissus and Parmenides, argue eristically (for their arguments assume falsehoods and are not deductive; or rather, the argument of Melissus is crude and contains no difficulty – grant him one absurdity (**a10**) and the others follow: this is not very hard). That Melissus draws fallacious inferences is clear: for he believes that, if everything that comes to be has a principle, then it is also assumed that everything that does not come to be does not have a principle. And this is absurd, too, that there is a principle of everything <that comes to be> – in the sense of a principle of the thing, not <just in the sense of a beginning> of the time <in which it comes to be> – and of generation, not of generation (**a15**) without qualification, but also of alteration, as if there were no change taking place at once. Furthermore, why should anything be unchanging if it is one? For as the part, too, being one, e.g. this water, moves in itself, why should the all not do the same? Furthermore, why should there not be alteration? Yet truly it is not possible that <what is altered> is one in form, unless this is meant with respect to that from which <alteration takes place> (a20) (and some of the philosophers of nature claim that <what comes to be> is one in this way, but not in that other way): for a human being is different in form from a horse and the opposites from each other.

Against Parmenides, too, one can use the same kind of arguments, even if there are some other arguments proper <to him>; and the solution is in one way that it is false, in another way that it does not entail the conclusion. It is false in as much as it assumes that being is said without qualification, (**a25**) when it is said in several ways; it does not entail the conclusion because, even if one assumed <the existence of> pale things only, given that the pale signifies one thing, pale things would nonetheless be many and not one. For the pale will be one neither by continuity nor in account: for being pale and what receives it will be different. And there will not be (**a30**) anything separate beyond the pale: for the pale and that to which it belongs are different not in as much as they are separate, but rather in being. However, Parmenides was not yet able to see this. In truth it is necessary to assume not only that being signifies one thing with reference to that of which it is predicated, but also that <it signifies> what is precisely being and what is precisely one. For the accident is said of some underlying thing, (**a35**) so that that, to which being belongs as an accident, will not be (for it is other than (**186b1**) being); therefore, there will be something which is not. So, what is precisely being will not belong to anything else. For it will not be possible to be it while being something determinate, unless being signifies several things in such a way that each of them is something. But it is assumed that being signifies one thing. If, then, what is precisely being does not belong as an accident to anything, (**b5**) but other things belong to it, why should it be the case that what is precisely being signifies what is rather than what is not? For if what is precisely being is also pale, but being pale is not being precisely being (it is not possible that being belong to it as an accident: for nothing which is not precisely being is being), it follows that the pale is not being: and not in the sense that it is not something determinate, but in the sense that it is not (**b10**) being at all. What is precisely being, then, is not: for it is true to say that it is pale, and the latter signifies what is not. So that the pale signifies what is precisely being, too: therefore, being signifies several things. Nor will what is have magnitude, if what is is what is precisely being: for to be is different for each of the parts.

It is also evident through this argument that what is precisely being divides into something else which is in turn precisely some being: (**b15**) for instance, if the human being is something which is precisely some being, it is necessary that also the animal and the biped be something which is precisely some being. For, if they are not something which is precisely some being, they will be accidents. Therefore either they will belong as accidents to the human being or to some other underlying thing. But this is impossible: what is said to be an accident is this, i.e. either what can belong or not belong, (**b20**) or that in whose account there belongs that to which it belongs (for instance, sitting is an accident that can be separated, while the account of the nose, to which we say that the snub belongs as an accident, belongs in <the account of> the snub). Furthermore <there are things> which are present in the definitional account or from which <the definitional account> is constituted, while (**b25**) the account of the whole does not belong in their account – for instance, the account of the human being does not belong in the biped or the account of the pale human being does not belong in the pale. If, then, these things are in this way and the biped belongs as an accident to the human being, it is necessary that either it is separable, so that it would be possible for the human being not to be biped, or (**b30**) the account of the human being will belong in the account of the biped. But this is impossible: for the latter belongs in the account of the former. But if the biped and the animal belong as accidents to something else and are not each precisely some being, then also the human being will be an accident of something else. But let it be established that what is precisely being is not an accident of anything, and that the composite of both is said of that of which both are said. (**b35**) Will it follow that the all is therefore made of indivisibles? (**187a1**) Some gave in to both arguments, namely to the argument according to which all things are one, if being signifies one thing, by conceding that what is not is, and to the argument from dichotomy, by introducing atomic magnitudes. But it is also evident that it is not true that, if being signifies one thing and it is not possible that (**a5**) contradictories be the case at the same time, there will be nothing which is not: for nothing prevents what is not from being some determinate thing which is not, without being without qualification. And truly it is absurd to say that, if there is nothing else beyond what is itself, all things will be one. For who understands what is itself if it is not as something that is precisely being? But, if so, then nothing prevents things that are from being many, (**a10**) as it has been said.

It is therefore clear that it is impossible that what is is one in this way;[10]

4. as for the physicists, on the other hand, there are two main ways in which they explain things. Accordingly, making what is a single body, i.e. what underlies – either one of the three, or something else which is more solid than fire but less solid than air – some (**a15**) produce everything else by means of density and rarity, and make [in this way] a plurality of things[11] (these are opposites, and, to put it in general terms, are excess and defect, as are, according to Plato, the great and the small, even though the one [i.e. Plato] makes the great and the small matter, and the one the form, while the others [i.e. the physicists] make the one, i.e. the underlying thing, matter, and make the opposites differences (**a20**) and forms). As for the others, <they argued that> the pairs of opposites are separated out from the one, in which they already were, just like Anaximander says, and also those who say that the things that are are both one and many,[12] just like Empedocles and Anaxagoras. For they too separate out everything else from the mixture. They differ, however, in that the former <argues that this process of separation> happens cyclically, while the latter (**a25**) <argues that> it happens once, and also in that the latter posits an unlimited number both of homeomerous bodies and of opposites, while the former posits only the so-called elements. It seems that Anaxagoras

[10] Using a semicolon instead of Ross's full stop after δῆλον, to keep as far as possible the unity of the μέν/δέ sentence.

[11] The text translated here is that transmitted by the Greek and Arabic tradition and is without Ross's correction. Following the text edited by Ross (οἱ μὲν γὰρ ἓν ποιήσανες τὸ [ὂν] σῶμα τὸ ὑποκείμενον) we would have: 'Accordingly, by making the underlying body one … some produce everything else by means of density and rarity.' On this point, see Cerami in this volume, section I.

[12] The text translated here is that transmitted by E, J^2 and the Greek-Arabic translation. Following the text transmitted by Simplicius and Philoponus, the first hand of the Vindoboniensis 100 (J) and two later manuscripts (F and I), Ross edits καὶ ὅσοι δ' ἓν καὶ πολλά φασιν εἶναι: 'and also those who say that <what underlies> is both one and many'.

made <homeomerous bodies and opposites> unlimited in this way because he accepted as true the general opinion of the physicists that nothing comes to be out of what is not (this is the reason why they claim 'things were all together' (**a30**), and why he makes coming to be a certain sort of thing alteration, while they talk about aggregation and separation). Moreover, <he made homeomerous bodies and opposites unlimited> also because opposites come to be out of one another; therefore, they must already have been present as constituents. For if everything which comes to be must do so either out of things that are or of things that are not, and since of these coming to be out of things that are not is impossible (all those who study nature (**a35**) agree on that opinion), they believe that the other possibility necessarily follows: everything comes to be out of things that are and that are present as constituents, but cannot be perceived by us in view of the smallness of their bulk. (**187b1**) This is the reason why they claim that everything is mixed in everything, because they saw everything coming to be from everything; <they claim that> things only appear different, and are said to be different from each other on the basis of what is most predominant, quantitatively, in the mixture of the unlimited things; as a matter of fact (**b5**), nothing is purely a whole pale, dark, sweet, flesh, or bone, but whichever each thing has most of seems to be its nature.

Now if the unlimited as unlimited is unknowable, then that which is unlimited in number or size will be a certain unknowable quantity, while that which is unlimited in form will be a certain unknowable quality. (**b10**) If, then, the principles are unlimited both in number and in form, there can be no knowledge of the things they make up. For we think we have knowledge of what is composite, when we know what things and how many it is from. Moreover, if it is necessary that if the part of a thing can be as large or small as you please, so can the whole thing (**b15**) (I mean one of the parts into which as constituents the whole is divided), and if it is impossible for an animal or a plant to be as large or small as you please, it is evident that neither can any of its parts be; for the whole would be in the same way. Now flesh and bone and such things are parts of animals, and fruits are parts of plants. (**b20**) It is, then, clearly impossible that flesh or bone or anything of that sort could proceed as far as you like in largeness or smallness. Further, if all such things are already present in one another, and do not come to be, but are merely separated out, being already there and are called according to what predominates in them, and if anything comes to be out of anything (for instance water is separated out of flesh, (**b25**) and flesh out of water), and if any limited body is exhausted by a limited body, it is evidently not possible that everything should be present in everything. For suppose that flesh has been removed from some water, and then more flesh came to be by separation from what remains: even if what is separated is lower each time, (**b30**) yet <the separated flesh> never goes below a certain quantity. Hence if separation comes to an end, it will not be the case that everything is in everything (there will be no more flesh in the remaining water); if, on the other hand, it does not come to an end, so that the extracting keeps going for ever, there will be an unlimited number of equal limited parts in a limited magnitude; but that is impossible. (**b35**) Besides, if every body must become smaller when something is removed from it, and if the quantity of flesh is defined both in magnitude and in smallness, it is evident that from the least possible quantity of flesh (**188a1**) no body can be separated. For there would be a quantity of flesh smaller than the least possible. Furthermore, an unlimited amount of flesh, blood and brain would be present in the unlimited bodies, separated from one another but nonetheless being things that are, and each unlimited; (**a5**) but that is nonsense. Although he [i.e. Anaxagoras] speaks in ignorance, he is right to say that the separating out will never be completed; for affections are not separable. If colours and states are mixed together, then if they are separated out, there will be a pale or healthy which is nothing else, not even <an attribute> of an underlying thing. Thus, in seeking impossible things, his Mind is absurd, if it wants (**a10**) to separate things, while it is impossible to do this either according to quantity or according to quality; according to quantity, since there is no smallest magnitude, according to quality, since affections are not separable. He does not get right even the coming to be of the things of the same form. For in one way clay breaks down into clay (a15) but in another it does not. And it is not in the same way that bricks are and come to be out of a house and a house out of bricks, and that water and air are and come to be out of each other. It is better to assume fewer and limited things <as principles>, as Empedocles does.

5. Now everyone makes the opposites principles – those who say that the universe is (**a20**) one and unchanging (for even Parmenides makes hot and cold principles, and these he calls fire and earth), or that the principles are rare and dense, or Democritus <who says that they are> the full and empty, of which he says that the one is as what is and the other is as what is not. Further <he says that things differ> by position, shape, arrangement: these are genera of opposites – of position they are above and (**a25**) below, in front and behind; of shape angled and unangular, straight and curved. That everyone in a way makes the principles opposites is, therefore, clear. And this is reasonable: for the principles must be neither from each other nor from other things, and all things must be from them. These things belong to the primary opposites – on the one hand because primary things are (**a30**) not from other things, and on the other because opposites are not from each other.

We must, however, investigate how this turns out on the basis of general considerations. It must, then, be grasped first that of all the things that are, nothing is of a nature to do or to undergo just anything under the agency of just anything; nor does anything whatever come to be from anything whatever – unless one takes things incidentally. (**a35**) For how could pale come to be from educated, unless the educated were an incidental to the not pale or the dark? But pale comes to be from not pale, and not from all of this, but (**188b1**) from dark or the things which are intermediate; and educated from not educated, only not from all but from uneducated or from something intermediate if there is such a thing. Nor, then, does anything cease to be into just anything first – for example, the pale does not cease to be into the educated – unless sometimes incidentally – (**b5**) but into the not pale, and not into just anything but into the dark or what is intermediate. And in the same way also the educated into the not educated, and this not into just anything, but into the uneducated or into something intermediate if there is such a thing. And this holds in the same way in the other cases, since (**b10**) the same account holds also for those of the things that are which are not simple but composite; but this is overlooked because the opposed conditions have not been named. For it is necessary for everything structured to come to be from unstructured, and the unstructured from structured, and for the structured to cease to be into unstructuredness – and this is not just any unstructuredness but the (**b15**) opposed one. It makes no difference whether we are speaking of structure or arrangement or composition; for it is clear that the same account applies. Indeed, house and statue and anything else whatever come to be in the same way: for the house comes to be from these things not being put together but being separated in this way, and the statue and any of the things which have been shaped <come to be> (**b20**) from shapelessness. And each of these is either an arrangement or some kind of composition. If this is true, therefore, everything which comes to be or which ceases to be would come to be or cease to be either from opposites or into opposites and the things which are intermediate between these; and the intermediate things are from the opposites – e.g. colours from pale and dark. Consequently (**b25**) all the things which come to be by nature would be opposites or from opposites.

Up to this point most other thinkers have been more or less in complete agreement, as we have said earlier. For even if they suppose things without giving an account, nevertheless everyone says that the elements and the things they call principles are the opposites – as if (**b30**) compelled by the truth itself. They differ from each other in that some take things that are prior and some things that are posterior, and some take things that are more knowable in account, as has been said, and some things that are more knowable by perception: for some suppose hot and cold to be causes of coming to be, some moist and dry, and others odd and even or strife and love, (**b35**) and these things differ in the way I have mentioned. Consequently in a way they say the same things and in a way different things from each other – they are different, as indeed they seem to most people to be, yet they are the same (**189a1**) by analogy. For they take them from the same column; for some of the opposites include others, and some are included by others. In this way, then, they speak in the same way and differently, and worse and better, and some take the things that are more knowable in account, as was said (**a5**) earlier, and others the things that are more knowable by perception (for the universal is knowable in account and the particular by perception; for the account is of the universal, while perception is of the individual) – for example the great and the small in account, and the rare and the dense by perception.

That the principles (**a10**) must be opposites is evident;[13]

6. what follows next is that we should say whether they are two or three or more. Well, they cannot be one, because the opposites are not one, and they cannot be unlimited, both because that which is[14] would not be knowable, and there is a single opposition in any single genus, and substance is some one genus, and because it is possible [to explain] on the basis of (**a15**) a limited number, and it is better [to explain] on the basis of a limited number – for instance as Empedocles does – rather than on the basis of an unlimited number; for he thinks that he explains just as much as Anaxagoras does on the basis of an unlimited number.

[13] Using a semicolon instead of Ross's full stop after φανερόν, to keep as far as possible the unity of the μέν/δέ sentence.

[14] To avoid confusion, the expression 'that which is', instead of our usual 'what is', is employed here to translate the phrase τὸ ὄν. This is because in this context the phrase 'what is' might mistakenly be understood as though it were the translation of a Greek expression such as ὃ ἔστιν, used elliptically for 'what is X', or 'what X is'. For a familiar example of this kind of usage, see *Republic* X, 597a2–4, c1, c9, where Plato employs 'what is' (ὃ ἔστιν) for the Form of Bed (ὃ ἔστι κλίνη), and contrasts 'what is' with something that is a bed (κλίνην τινά). The argument at 189a13 is not about the knowability of 'what is' in this sense. Aristotle is not talking about being, or what being is, or (for some X) what X is, but rather is speaking generically about the things that are (τὰ ὄντα). He presupposes that knowledge of τὰ ὄντα is based upon their principles, and that the very possibility of such principle-based knowledge of the things that are would be undermined should their principles be unlimited.

Furthermore, some opposites are prior to others, and some come to be from others; for instance, sweet and bitter and pale and dark, but the principles must always remain.

Well, while it is clear from these points that they are neither one nor an unlimited number, given that they are limited in number, there is some reason for not making them only two; for one might raise a difficulty about how either density is of such a nature as to make rarity something, or rarity make density something; likewise too for any other opposition, for Love does not gather Strife and make something from it, nor does Strife make something from it, but both <make> some other third thing <into something>. And some assume even more things from which they construct the nature of the things that are. In addition to these points, one might raise a further difficulty over this, should one not posit another nature for the opposites: we do not see the opposites to be the substance of any of the things that are, and the principle ought not be said of some underlying thing. For <if so> there will be a principle of a principle, for what underlies is a principle, and is thought to be prior to what is predicated. Further, we say that a substance is not opposite to a substance. Well then, how could a substance be <constituted> from non-substances?; or how could what is not a substance be prior to a substance? Consequently, should one hold that both the earlier account and this account are true, it is necessary, if one intends to preserve them both, to posit some third thing, just as those saying that the universe is some single nature affirm; <saying that it is>, for example, water or fire or what is intermediate between them; but it is thought that what is intermediate is more so, for fire and earth and air and water are already entangled with oppositions. Consequently, both those who make the underlying thing different from these [four], and of the others, those who make <it> air, do so not without reason; for of the others, air too has the least perceptible differentiae; and what follows next is water. But at any rate all of them shape this one [underlying thing] by means of the opposites, by density and rarity, and by the more and less; and these are clearly in general excess and deficiency, just as was said earlier. Also, this opinion, that the one and excess and deficiency are principles of the things that are, seems to be old, though not in the same way, but rather the earliest thinkers say the two act and the one is acted upon, whereas some of the later thinkers say rather the opposite, that the one acts and the two are acted upon.

Well, there would seem to be some reason to assert that the elements are three for those who are looking closely on the basis of these and other such considerations, just as we said, but that they are more than three no longer <would there be>. For the one [underlying thing] is sufficient to be acted upon, and if being four <opposites> (**b20**) there will be two oppositions, some other intermediate nature will need to belong separately to each; but if they are able to generate from one another, being two <oppositions>, one of the oppositions would be superfluous. At the same time, it is also impossible for there to be many primary oppositions. For substance is some single genus of what is, so that the principles will differ (**b25**) from one another only in being prior and posterior, but not in genus; for in every case there is a single opposition in the case of a single genus, and all oppositions are thought to be led up to a single <opposition>. Well, then, that there is neither one element, nor more than two or three is clear; but whether there are two or three, just as we have said, is very difficult.

7. (**b30**) This then is what we ourselves say, beginning with a discussion of coming to be in its entirety. It is natural first to speak of the common features and on this basis to examine what is distinctive about particular cases. We say that one thing comes to be from another and that something comes to be from something different, when speaking of simples or compounds. What I mean is this: it is possible for a human being to come to be (**b35**) educated, for that which is not educated to come to be educated or for (**190a1**) a non-educated human to come to be an educated human. By a simple I mean when the human and what is not educated is what is coming to be, and when the not educated is what comes to be. What comes to be and what is coming to be are compounded when we say that the not educated human (**a5**) comes to be an educated human. In some of these cases we say not only that this is coming to be but also that it is coming to be from this (as when we say that the educated comes from what is not educated) but this is not said in all cases. For it is not from a human that something educated came to be but a human came to be educated. Of the things that come to be, in cases we describe as involving simples, in one case something (**a10**) comes to be by remaining, in the other by not remaining. For the human remains when coming to be an educated human and is an educated human but that which is not educated and the uneducated remains neither without qualification nor in a composite form.

In the light of these distinctions, one can grasp in all cases of things that come to be, if one looks closely at the phenomena in the manner in which we speak of them, that something must (**a15**) always underlie (i.e. that which is in the process of coming to be) and that this [i.e. that which is in the process of coming to be], even if it is one in number, is not one in form (by 'in form' I mean 'in account'). For being a human is not the same as being uneducated. And the one remains, the other does not. That which is not opposed remains (the human remains), the not educated and the (**a20**) uneducated does not, nor that which is put together out of these (e.g. the uneducated human). The expression 'this comes to be from that' (as opposed to the expression 'this comes to be that') is more used in the case of what does not remain, as when we say that what is educated comes to be from what is uneducated but not that it [i.e. the educated] comes to be from human. Nonetheless we do also sometimes speak in a similar way [i.e. using the 'this comes to be from that . . .' way] about what remains, (**a25**) saying that from bronze a statue comes to be, not the bronze comes to be a statue. However, we do speak in both ways in the case of something opposed that does not remain: we say both that this comes to be from that and this comes to be that. And so we talk of something educated coming from something which is not educated and of something which is not educated coming to be something which is educated. This is why we speak in a similar way about what is put together. For we say that both: from an uneducated (**a30**) human and the uneducated human comes to be educated. 'Coming to be' is said in many ways, and some things are said not to come to be but to come to be something or other, but substances alone are said to come to be without qualification. In all cases other than <the coming to be of substance>, it is evident that there must be something underlying, that which comes to be. For when a thing comes to be of such and such quantity or quality or (**a35**) in a given relation, at a given time or place, something is always present underlying (since substance alone is not said of another underlying thing, (**190b1**) while everything else is said of substance). But that even substances, and whatever things simply are, come to be from something underlying will be evident to one who looks closely. For always there is something that underlies, from which the thing generated comes to be: as in the case of plants and animals coming to be from (**b5**) a seed. Some of the things that come to be in the unqualified way come to be some by change of shape (as a statue), others by addition (as things that grow bigger), others by subtraction (as Hermes from the stone), others by composition (as a house), others by alteration (as things changed in accordance with their matter). It is evident that all things that come to be in this way (**b10**) come to be from what underlies. From what has been said it is clear that everything that comes to be is always a composite: there is something that comes to be and something that comes to be that <thing which comes to be>. There are two types of the latter: either that which underlies or what is opposed <to that which comes to

be>. I say that the uneducated is opposed and human underlies; and (**b15**) shapelessness and formlessness and lack of order are opposed and bronze or stone or gold is what underlies.

It is evident that if there are indeed causes and principles of things which are by nature – from which first causes and principles these things are and come to be non-incidentally what it is said to be in describing its nature [*ousia*], then (**b20**) everything comes to be both from what underlies and its shape. For an educated human is, in some way, composed of a human and educated, in that it is decomposable into the accounts of these. It is clear, therefore, that things that come to be do so from these. What underlies is one in number but two in form. The (**b25**) human, the gold and in general the matter which is countable <is one in number>. It is more a *this such* [i.e. a determinate particular] and the thing that comes to be does not come to be from it incidentally. The privation and the opposition is incidentally. The form is one: for example, the arrangement or being educated, or any of the things predicated in this way. This is why it is the case that we should say, on the one hand, that the principles are two, (**b30**) on the other that they are three. For there are two as opposites (as when one speaks of the educated and the uneducated or the hot and the cold or the harmonized and the disharmonized), but in a way this is not so. For the opposites cannot be acted upon by each other. The problem is solved because what underlies is different from these: it is not (**b35**) an opposite. So in one way there are no more principles than the opposites, but they are so to say two in number; but they are not entirely two, (**191a1**) but three since something else is present which is essentially different from them. For being a human and being uneducated differ as do being shapeless and being bronze.

We have said how many principles there are for the coming to be of natural things and in what way there are that many. It is clear that there must be something which underlies (**a5**) the opposites and that the opposites are two <in number>. But, in another way, it is not necessary <to have three principles>: for one or other of the opposites will be enough to make the change by its absence or presence. The underlying nature is knowable by analogy. For as bronze stands to a statue or wood to a bed or (**a10**) the matter – that is, what is shapeless before it takes on shape – stands to what is shaped, so the underlying nature stands to the substance, which is a this such [i.e. an informed object] and what is. This then is one principle, although it is not one in the way a this such is one nor does it exist in the way a this such exists; there is also one other principle: that of which there is the account, and furthermore another: what is opposite to it, the privation. We have stated above in what way these are two principles, in what way (**a15**) there are more. For – to resume – it was said first that the opposites alone are the principles but later that there must be something that underlies and so that there are three principles. From these considerations it is now evident what distinguishes the opposites, how the principles are related to each other, and what type of thing that which underlies is. Whether the form or (**a20**) that which underlies is substance is not yet clear. But it is clear that there are three principles, in what way there are three, and what kind of principle each is. This completes our examination of how many the principles are and what they are.

8. After these things let us say that the difficulty faced by the earliest thinkers, too, is solved only in this way. For those who were the first to inquire (**a25**) philosophically into truth and the nature of the things that are were led astray by their inexperience,[15] which as it were thrust them onto another road. So they say that none of the things that are either comes to be or ceases to be, because it is necessary that what comes to be does so either from what is or from what is not; but from both of these it is impossible. (**a30**) For what is cannot come to be (because it already is), and from what is not nothing could come to be (because something has to underlie). And so, overplaying the immediate consequence [of the difficulty about coming to be], they say that the many things do not exist, but only what is itself exists.

They acquired this opinion for the reasons which have been stated. We, however, say that, in one way, to come to be from what is, (**a35**) or from what is not, or for what is not or what is to do something, or to be acted upon, or come to be a this, of any sort, is no different (**191b1**) from the physician's doing something, or being acted upon, or something's being from the physician, or <something's> coming to be <from the physician>, and so as this <latter> is said in two ways, it is clear that also <being or coming to be> from what is, and what is doing something, or being acted upon, too <are said in two ways>. A physician builds a house, not in so far as he is a physician, but in so far as (**b5**) he is a housebuilder, and he comes to be pale, not in so far as he is a physician, but in so far as he is dark. On the other hand, he heals or fails to be a physician in so far as he is a physician. Since we say most properly that the physician does

[15] *OCT* (F, I, V, Philoponus, Themistius): ἀπειρίας. E, J, Simplicius: ἀπορίας. This latter version would mean: '... were led astray by this difficulty'.

something, or is acted upon, or <something> comes to be from a physician, when these are acted upon him, or he does <these>, or comes to be <these> in so far as he is a physician, it is clear that also 'to come to be from what is not' signifies '<to come to be from what is not> (**b10**) in so far as it is what is not'. They did not make this distinction and gave up on this issue; and it is through this error that they went so much further astray as to suppose that none of the other things comes to be or is, but did away with all coming to be. We too, on our part, say that nothing comes to be without qualification from what is not. But nevertheless <something> comes to be in some way from what is not – that is, (**b15**) incidentally. (For from the privation, which is in itself what is not – an item that is not present as a constituent in <what comes to be>, something comes to be. Yet this is a cause of surprise, and it seems impossible that something should come to be this way, from what is not.) In the same way <we maintain> that <something> does not come to be from what is, and what is does not come to be <something>, except incidentally. This, too, comes to be this way – the same way as if, e.g. from animal animal (**b20**) might come to be, and from some animal some animal <might come to be>. Thus, suppose a dog to come to be from a horse.[16] The dog would not only come to be from some animal, but also from animal, but not in so far as it is an animal, for that is already present. But if something is going to come to be an animal, not incidentally, it will not be from animal: and if <something is going to come to be> what is, <not incidentally, it will> not <be> from (**b25**) what is. Not from what is not either, for we have said what 'from what is not' signifies, namely that <from what is not > in so far as it is what is not. Furthermore we do not do away with the principle that everything either is or is not.

This then is one way [of solving the difficulty]. Another is that the same things can be said according to the potentiality and according to the actuality. This has been explained with greater precision elsewhere. (**b30**) So, just as we said, the difficulties which forced people to do away with some of the things we mentioned are now solved. For it was through this that the earliest thinkers also were thrust away so much from the road that leads to coming to be and ceasing to be, and, in general, change. For this very nature, had they come to see it, would have solved all their errors.

9. (**b35**) There are certain others who have touched on this, though not sufficiently. For first, they agree that there is coming to be without

[16] *OCT* follows Laas in reading 'suppose a dog to come to be <from a dog, or a horse> from a horse'.

qualification from what is not,[17] (**192a1**) in which respect <they agree that> Parmenides spoke correctly. Second, it is apparent to them that if <what underlies> is one in number, it is also only one potentially – but this is a very different thing. For we ourselves say that matter and privation are different, and that of these one, the matter, is what is not incidentally, (**a5**) while privation is in itself what is not; and that the one, the matter, is near substantial being and is substantial being in a way, while the other is in no way at all substantial being; but these other people <say that> the great and the small are alike what is not, either the two taken together or each one separately.[18] So this type of triad and the previous one <of ours> are completely different. They advanced to the point <of recognizing> that (**a10**) there needs to be a certain nature to underlie, yet they make this one. For even if someone makes it a dyad, calling it great and small, nonetheless it comes to the same, for he overlooks the other nature. The nature which remains is a joint cause with the shape of things that come to be, as a mother; the other part of the opposition might often (**a15**) be imagined not to exist at all, if one directed one's thought to its maleficence. For, supposing there is a certain thing that is divine, good and desirable, we say there to be one thing that is its opposite, and another which by nature desires it and reaches out for it in accordance with its nature. For them, on the other hand, it results that the opposite reaches out for (**a20**) its own ceasing to be. And yet neither is the form able to desire itself, on account of not being in need, nor is the opposite (for opposites destroy each other), but this is the matter, just as if the female reached out for the male and the ugly for the beautiful – except that in itself it is neither ugly nor female, but only (**a25**) incidentally so.

In a way it is possible that what underlies ceases to be and comes to be, and in a way not. Considered as that in which, it ceases to be in its own right (for that which ceases to be, the privation, is in this); but considered as potential, it does not cease to be in its own right, but it is necessary that it be incapable of ceasing to be or coming to be. For if it came to be, there would need to be something prior to underlie from which, (**a30**) as a constituent, <it came to be>; but this is its nature,[19] so it will be before it comes to be (for by matter I mean the primary underlying subject for each thing, from which, as a constituent, something comes to be non-incidentally). And if it

[17] Omitting the τι, which is absent from the best mss.

[18] Taking 'the great and the small' as the grammatical subject, on the supposition that the discussion at this point concerns what should be called 'what is not' and in what way it should be so called.

[19] Reading αὐτῆς with IP.

ceased to be it would finally come to this, so it will have ceased to be before it had ceased to be. Concerning the principle according to the form, whether it is (**a35**) one or <they are> many and what it is or what they are, it is the work of first philosophy to determine with precision – so let us put <such questions> aside until that occasion. (**192b1**) In the following discussions we will be speaking about the forms of natural and perishable things.[20] So, then, let us take it as determined in this manner that there are principles and which and how many in number they are; but let us carry on the discussion, beginning from another starting point.

[20] Ross records καὶ τῶν φθαρτῶν as the reading of the Λ mss. (FIJ), followed by Bekker. See Lennox 2008: 153 n. 4 for his defence of the text and translation provided here.

References

Ackrill, J. L. 1963. *Aristotle's Categories and De Interpretatione: Translated with Notes*, Oxford: Oxford University Press

Anagnostopoulos, A. 2013. 'Aristotle's Parmenidean Dilemma', *Archiv für Geschichte der Philosophie* 95: 245–74

Aquinas, T. 1963. *Commentary on Aristotle's Physics*, trans. R. J. Blackwell, R. J. Spath and W. E. Thirkel, New Haven: Yale University Press

1985. *Sententia libri De sensu et sensato, cuius secundus tractatus est De memoria et reminiscencia* (Sancti Thomae de Aquino Opera Omnia 45, 2), Rome and Paris: Vrin

Averroes. 1962. *Commentarium Magnum in Aristotelis* De Physico *Audito libri octo*, in *Aristotelis Opera cum Averrois Commentariis*, Frankfurt am Main: Minerva, 14 vols. [Venetiis: Junctas, 1562–74], vol. IV

Badawī, A. (ed.) 1964–5. *Arisṭūṭālīs. Al-Ṭabīʿa. Tarǧamat Isḥāq Ibn Ḥunayn maʿa šurūḥ Ibn al-Samḥ wa Ibn ʿAdī wa Mattā Ibn Yūnus wa Abū al-Faraǧ Ibn al-Ṭayyib*, Cairo: al-Dār al-Qawmiyya li al-Ṭibāʿa wa al-Našr, 2 vols.

Baltussen, H., Mueller I., Atkinson M. and Share M. 2012. *Simplicius on Aristotle Physics I. 5–9*, London: Bloomsbury

Barnes, J. 1979. 'Parmenides and the Eleatic One', *Archiv für Geschichte der Philosophie* 61: 1–21

1994. *Aristotle, Posterior Analytics*, 2nd edn, Oxford: Oxford University Press

1995. 'Metaphysics', in *The Cambridge Companion to Aristotle*, ed. J. Barnes, Cambridge: Cambridge University Press: 66–108

Barney, R. 2012. 'History and dialectic (*Metaphysics* A 3, 983a24–984b8)', in *Aristotle's Metaphysics Alpha: Symposium Aristotelicum*, ed. C. Steel, Oxford: Oxford University Press: 69–104

Berti, E. 1969. 'Physique et métaphysique selon Aristote: *Phys.* I 2, 184b25–185a5', in *Naturphilosophie bei Aristoteles und Theophrast*, ed. I. Düring, Heidelberg: Lothar Stiehm Verlag: 18–31

Bogen, J. 1991. 'Aristotelian contrariety', *Topoi* 10: 53–66

1992. 'Contrariety and change in Aristotle', *Phronesis* 37: 1–21

Bogen, J. and McGuire, J. E. 2006. 'Aristotle on motion, change, and predication', in *Essays in Ancient Greek Philosophy*, eds. R. N. Bosley and C. Y. Panayides, Nicosia: University of Nicosia Press: 1–32

Bolton, R. 1991. 'Aristotle's method in natural science: *Physics* I', in *Aristotle's Physics: A Collection of Essays*, ed. L. Judson, Oxford: Clarendon Press: 1–29

Bostock, D. 1982. 'Aristotle on the principles of change', in *Language and Logos: Studies in Ancient Greek Philosophy Presented to G. E. L. Owen*, eds. M. Schofield and M. C. Nussbaum, Cambridge: Cambridge University Press: 179–96

2006a. 'Aristotle on the principles of change in *Physics* I', in *Space, Time, Matter and Form: Essays on Aristotle's Physics*, ed. D. Bostock, Oxford: Oxford University Press: 1–18

2006b. 'Aristotle on the Eleatics in *Physics* I.2–3', in *Space, Time, Matter, and Form: Essays on Aristotle's Physics*, ed. D. Bostock, Oxford: Oxford University Press: 103–15

Broadie, S. (forthcoming). 'Physics I.9', in *Aristotle's Physics I, Symposium Aristotelicum*, eds. K. Ierodiakonou, P. Kalligas and V. Karasmanis, Oxford: Oxford University Press

Brunschwig, J. 2004. '*On Generation and Corruption* I.1: a false start?', in *Aristotle: On Generation and Corruption, Book I: Symposium Aristotelicum*, eds. F. De Haas and J. Mansfeld, Oxford: Oxford University Press: 25–63

Burnyeat, M. 2001. *A Map of Metaphysics Zeta*, Pittsburgh: Mathesis Publications

2011. 'Episteme', in *Episteme etc.: Essays in Honour of Jonathan Barnes*, eds. B. Morison and K. Ierodiakonou, Oxford: Oxford University Press: 3–29

Castelli, L. M. 2010. *Problems and Paradigms of Unity: Aristotle's Accounts of the One*, Sankt Augustin: Academia Verlag

2018. *Aristotle, Metaphysics I: A Translation and Commentary*, Oxford: Oxford University Press

Cerami, C. 2005. 'La sostanza sensibile e la nozione di *tode toionde* in *Metafisica* VII 8', in *Eidos-Idea: Platone, Aristotele e la tradizione platonica*, eds. F. Fronterotta and W. Leszl, Sankt Augustin: Academia Verlag: 211–32

2009. 'Thomas D'Aquin lecteur critique du *Grand Commentaire* d'Averroès à *Phys.* I,1', *Arabic Sciences and Philosophy* 19: 189–223

2012. 'Mélange: *Minima Naturalia* et croissance animale dans le *Commentaire moyen* d'Averroès à *De Generatione et Corruption* I,5, in *La nature et le vide dans la physique médiévale*, eds. J. Biard and S. Rommevaux, Turnhout: Brepols: 137–64

2015. *Génération et substance: Aristote et Averroès entre physique et métaphysique*, Berlin: W. de Gruyter

Charles, D. 2003. 'Prime matter and elemental change', in *Aristotle: On Generation and Corruption, Book I: Symposium Aristotelicum*, eds. F. de Haas and J. Mansfeld, Oxford: Oxford University Press: 151–69

2008. 'Aristotle's psychological theory', *Proceedings of the Boston Area Colloquium* 24: 1–30

2010. '*Metaphysics* Θ. 7 and 8: some issues concerning actuality and potentiality', in *Being, Nature, and Life in Aristotle*, eds. J. G. Lennox and R. Bolton, Cambridge: Cambridge University Press: 168–97

Charlton, W. 1970. *Aristotle: Physics Books I and II, Translated with Introduction and Notes*, Oxford: Clarendon Press

1992. *Aristotle: Physics Books I and II*, Oxford: Clarendon Press

Cherniss, H. 1935. *Aristotle's Criticism of Presocratic Philosophy*, Baltimore: Johns Hopkins University Press

1944. *Aristotle's Criticism of Plato and the Academy*, Baltimore: Johns Hopkins University Press

Clarke, T. 2015. 'Aristotle and the ancient puzzle about coming to be', *Oxford Studies in Ancient Philosophy* 49: 129–50

Code, A. 1976a. 'The persistence of Aristotelian matter', *Philosophical Studies* 93: 357–67

1976b. 'Aristotle's response to Quine's objections to modal logic', *Journal of Philosophical Logic* 5: 159–86

Coxon, A. H. 2009. *The Fragments of Parmenides: A Critical Text with Introduction, Translation, the Ancient Testimonia and a Commentary*, rev. and expanded edn., Las Vegas: Parmenides Publishing

Crubellier, M. (forthcoming). 'Physics I.2', in *Aristotle's Physics I, Symposium Aristotelicum*, eds. K. Ierodiakonou, P. Kalligas and V. Karasmanis, Oxford: Oxford University Press

Curd, P. 2004. *The Legacy of Parmenides: Eleatic Monism and Later Presocratic Thought*, rev. paperback edn, Las Vegas: Parmenides Publishing [Princeton: Princeton University Press, 1998]

de Haas, F. 1998. 'Philoponus on Theophrastus on composition in nature', in *Theophrastus: Reappraising the Sources*, eds. J. M. van Ophuijsen and M. van Raalte (Rutgers University Studies in Classical Humanities), New Brunswick, NJ, and London: Transaction: 171–89.

Diels, H. 1912. *Die Fragmente der Vorsokratiker, griechisch und deutsch,* Berlin: Weidmann

Dillon, J. M. 2003. *The Heirs of Plato: A Study of the Old Academy*, Oxford: Clarendon Press

Edwards, M. J. 1994. *Philoponus on Aristotle's Physics 3*, Ithaca: Cornell University Press

Falcon, A. 2005. *Aristotle and the Science of Nature: Unity without Uniformity*, Cambridge: Cambridge University Press

2012. 'The reception of Aristotle's study of animal motion in the Latin world', *Documenti e Studi sulla Tradizione Filosofica Medievale* 22: 521–39

2016. 'The subject matter of Aristotle's *Physics*', in *Soma: Körperkonzepte und körperliche Existenz in der antiken Philosophie und Literatur*, eds. T. Buchheim, D. Meissner and N. Wachsmann, Hamburg: Felix Mainer Verlag: 423–36

Ferguson, J. 1969. 'The opposites', *Apeiron* 3: 1–17

Frede, M. 1987. 'The unity of general and special metaphysics: Aristotle's conception of metaphysics', in M. Frede, *Essays in Ancient Philosophy*, Minneapolis: Minnesota University Press: 81–95

Furley, D. J. 1967. *Two Studies in the Greek Atomists*, Princeton: Princeton University Press

1989. 'Anaxagoras in Response to Parmenides', in D. J. Furley, *Cosmic Problems*, Cambridge: Cambridge University Press: 47–65

1992. 'From Anaxagoras to Socrates', in *The Philosophy of Socrates,* vol. II: *Elenchus, Ethics, and Truth*, ed. K. Boudouris, Athens: International Center for Greek Philosophy and Culture: 74–80

2002. 'Anaxagoras, Plato, and the Naming of Parts', in *Presocratic Philosophy: Festschrift for A. P. D. Mourelatos*, eds. V. Caston and D. Graham, Aldershot: Ashgate: 119–26

Gershenson, D. and Greenberg, D. 1962. 'Aristotle confronts the Eleatics: two arguments on "the One"', *Phronesis* 7: 137–51

Giardina, G. 2002. *I fondamenti della fisica: analisi critica di Aristotele, Phys. I*, Catania: CUEM

Gill, M. L. 2012. *Philosophos: Plato's Missing Dialogue*, Oxford: Oxford University Press

Gotthelf, A. 2012. *Teleology, First Principles, and Scientific Method in Aristotle's Biology*, Oxford: Oxford University Press

Graham, D. W. 1994. 'The postulates of Anaxagoras', *Apeiron* 27: 77–121

Gregoric, P. 2006. 'Quantities and contraries: Aristotle's *Categories* 6, 5b11–6a18', *Apeiron* 39: 341–58

Haack, S. 1978. *Philosophy of Logics*, Cambridge: Cambridge University Press

Hackforth, R. 1972. *Plato's Philebus*, Cambridge: Cambridge University Press

Hardie, R. P. and Gaye, R. K. 1930. '*Physica*', in *The Works of Aristotle Translated into English*, ed. W. D. Ross, vol. II, Oxford: Oxford University Press

Harriman, B. 2015. 'The beginning of Melissus' *On Nature, Or On What-Is*: a reconstruction', *Journal of Hellenic Studies* 135: 19–34

Horstschäfer, T. M. 1998. '*Über Prinzipien': Eine Untersuchung zur methodischen und inhaltlichen Geschlossenheit des ersten Buches der Physik des Aristoteles*, Berlin and New York: W. de Gruyter

Husik, I. 1952. *Philosophical Essays: Ancient, Medieval and Modern*, eds. M. C. Nahm and L. Strauss, Oxford: Blackwell

Ibn al-Nadīm, 1871–2. *Kitāb al-Fihrist, mit Anmerkungen herausgegeben von G. Flügel, nach dessen Tode besorgt von J. Roediger und A. Müller*, Leipzig: Vogel, 2 vols.

1970. *The Fihrist of al-Nadīm: A Tenth-Century Survey of Muslim Culture*, trans. B. Dodge, New York and London: Columbia University Press, 2 vols.

Inwood, B. and Wolf, R. 2013. *Aristotle: Eudemian Ethics* (Cambridge Texts in the History of Philosophy), Cambridge: Cambridge University Press

Irwin, T. H. 1988. *Aristotle's First Principles*, Oxford: Oxford University Press

Jones, B. 1974. 'Aristotle's introduction of matter', *Philosophical Review* 83: 474–500

Judson, L. (forthcoming). *Aristotle: Metaphysics Λ: A Translation and Commentary*, Oxford: Oxford University Press

Katz, E. C. 2008. 'A Question of Numbers: Aristotle's Critique of Platonic Mathematical Objects, Form-Numbers, and First Principles in

"Metaphysics" M-N', available from ProQuest Dissertations & Theses Global: Order No. 3303019

Kelsey, S. 2006. 'Aristotle *Physics* I 8', *Phronesis* 51: 330–61

2008. 'The place of I 7 in the argument of *Physics* I', *Phronesis* 53: 180–208

2010. 'Hylomorphism in Aristotle's *Physics*', *Ancient Philosophy* 30: 107–24

2015. 'Aristotle on interpreting nature', in *Aristotle's Physics: A Critical Guide*, ed. M. Leunissen, Cambridge: Cambridge University Press: 31–45

Kenny, A. 2011. *Aristotle, The Eudemian Ethics: A New Translation*, Oxford: Oxford University Press

Keynes, J. N. 1894. *Studies and Exercises in Formal Logic*, 3rd edn, London and New York: Macmillan

Laas, E. 1836. *Aristotelische Textes-Studien*, Berlin: Reimer

Laks, A. 1998. 'Le début d'une physique: ordre, extension et nature des fragments 142/144 A/B de Théophraste', in *Theophrastus: Reappraising the Sources*, eds. J. M. van Ophuijsen and M. van Raalte (Rutgers University Studies in Classical Humanities), New Brunswick, NJ, and London: Transaction: 143–63

Lederman, H. 2014. '*ho pote on esti* and coupled entities: a form of explanation in Aristotle's natural philosophy', *Oxford Studies in Ancient Philosophy* 46: 109–64

Lefebvre, D. 2008. 'L'usage de la série des contraires en *Métaphysique* Γ 2, 1004b27–1005a6', in *Aristote: Métaphysique Gamma*, eds. M. Hecquet-Devienne and A. Stevens, Louvain-la-Neuve: Peeters: 287–321

Lemmon, E. J. 1965. *Beginning Logic*, London: Nelson

Lennox, J. G. 2008. '"As if we were investigating snubness": Aristotle on the prospects for a single science of Nature', *Oxford Studies in Ancient Philosophy* 35: 149–86

Lettinck, P. 1994. 'Problems in Aristotle's *Physics* I, 1 and their discussion by Arab commentators', *Journal for the History of Arabic Sciences* 10: 89–109

Lewis, F. A. 2008. 'What's the matter with prime matter?', *Oxford Studies in Ancient Philosophy* 34: 123–46

2013. *How Aristotle Gets By in Metaphysics Zeta*, Oxford: Oxford University Press

Lloyd, G. E. R. 1966. *Polarity and Analogy: Two Types of Argumentation in Early Greek Thought*, Cambridge: Cambridge University Press

Loux, M. J. 1992. 'Aristotle and Parmenides: an interpretation of *Physics* A8', *Boston Area Colloquium in Ancient Philosophy* 8: 281–319

Mansfeld, J. 1986. 'Aristotle, Plato, and the Preplatonic doxography and chronology', in *Storiografia e dossografia nella filosofia antica*, ed. G. Cambiano, Turin: Tirrenia: 1–59

Mansion, A. 1945. *Introduction à la Physique aristotélicienne*, 2nd edn, Louvain: Éditions de l'institut supérieur

Mansion, S. 1953. 'Aristote, critique des Éléates', *Revue philosophique de Louvain* 51: 165–86

ed. 1961. *Aristotle et les problèmes de méthode*, Louvain and Paris: Publications universitaires de Louvain et éditions Béatrice–Neuwelaerts

1984. 'Le role de l'exposé et de la critique des philosophies antèrieures chez Aristote', in *Études aristotéliciennes*, ed. S. Mansion, Louvain: Éditions de l'institut supérieur: 55–76

Mates, B. 1965. *Elementary Logic*, Oxford: Oxford University Press

Matthews, G. B. 1972. 'Senses and kinds', *Journal of Philosophy* 69: 149–57

Menn, S. 1995. 'Metaphysics, dialectic and the *Categories*', *Revue de Métaphysique et de Morale* 100: 311–37

(n.d.) '*Physics* II and its relation to *Physics* I–IV', unpublished paper

(forthcoming). 'Physics I.1', in *Aristotle's Physics I, Symposium Aristotelicum*, eds. K. Ierodiakonou, P. Kalligas and V. Karasmanis, Oxford: Oxford University Press

Mignucci, M. 2007. 'Aristotle on the existential import of propositions', *Phronesis* 52: 121–38

Mourelatos, A. P. D. 2008. *The Route of Parmenides*, rev. and expanded edn, Las Vegas: Parmenides Publishing

Osborne, C. 2009. *Philoponus on Aristotle's Physics I. 4–9*, London: Duckworth

Owen, G. E. L. 1960. 'Eleatic questions', *Classical Quarterly* 10: 84–102

Palmer, J. 2003. 'On the alleged incorporeality of what is in Melissus', *Ancient Philosophy* 23: 1–10

2004. 'Melissus and Parmenides', *Oxford Studies in Ancient Philosophy* 26: 19–54

2009. *Parmenides and Presocratic Philosophy*, Oxford: Oxford University Press

Pellegrin, P. 1994. 'Aristote et la physique éléate', in *Passion des formes: dynamique qualitative, sémiophysique et intelligibilité*, ed. R. Thom, Fontenay-St. Cloud: ENS Éditions: 121–37

2002. *Aristote, Physique*, 2nd edn, Paris: Flammarion

Peramatzis, M. 2011. *Priority in Aristotle's Metaphysics*, Oxford: Oxford University Press

Peters, F. E. 1968. *Aristoteles Arabus*, Leiden: Brill

Quarantotto, D. 2005. *Causa finale, sostanza, essenza in Aristotele*, Naples: Bibliopolis

2016. 'Aristotle's way away from Parmenides' way: a case of scientific controversy and ancient humour', *Elenchos* 36: 209–28

2017. *L'Universo senza spazio. Aristotele e la teoria del luogo*, Naples: Bibliopolis

(forthcoming). 'Physics I 3', in *Aristotle's Physics I, Symposium Aristotelicum*, eds. K. Ierodiakonou, P. Kalligas and V. Karasmanis, Oxford: Oxford University Press

Rashed, M. 2005. *Aristote: De la génération et la corruption*, Paris: Les Belles Lettres

Ross, W. D. 1936. *Aristotle. Physics: A Revised Text with Introduction and Commentary*, Oxford: Clarendon Press

1951. *Plato's Theory of Ideas*, Oxford: Clarendon Press

Rossi, G. 2006. 'Desanuadando argomentos: las aplicaciones filosoficas de la dialéctica segùn las Refutaciones Sofisticas', *Méthexis* 19: 79–109

2014. 'Aristotle on untying arguments', in *Zur modernen Deutung der aristotelischen Logik*, eds. N. Öffenberger and A. Vigo, Hildesheim: Olms, vol. XI: 105–40

Rowe, C. J. 1999. *Plato: Statesman*, Indianapolis and Cambridge, MA: Hackett
Sayre, K. 2005. *Plato's Late Ontology: A Riddle Resolved*, Las Vegas: Parmenides Publishing
Schofield, M. 1980. *An Essay on Anaxagoras*, Cambridge: Cambridge University Press
Schramm, M. 1962. *Die Bedeutung der Bewegungslehre des Aristoteles für seine beiden Lösungen der zenonischen Paradoxie*, Frankfurt am Main: Klostermann
Sharples, R. W. 1992. *Alexander of Aphrodisias, Quaestiones 1.1–2.15*, London: Duckworth
1998. *Theophrastus of Eresus, Commentary, Vol. 3.1: Sources on Physics*, Leiden and New York: Brill
Simplicius, 1882. *In Aristotelis Physica*, ed. H. Diels, Berlin: Reimer
Simpson, P. L. P. 2013. *The Eudemian Ethics of Aristotle*, New Brunswick, NJ, and London: Transaction
Sorabji, R. 2012. 'Introduction', in *Simplicius: On Aristotle Physics 1.5–9*, trans. H. Baltussen, M. Atkinson, M. Share and I. Mueller, London: Bloomsbury
Stevens, A. 2012. *Aristote, La physique, Nouvelle traduction annotée*, Paris: Vrin
Stokes, M. C. 1971. *One and Many in Presocratic Philosophy*, Washington, DC: Center for Hellenic Studies
Tannery, P. 1930. *Pour l'histoire de la science hellène*, 2nd edn, Paris: Gauthier-Villars
Tarán, L. 1965. *Parmenides: A Text with Translation, Commentary, and Critical Essays*, Princeton: Princeton University Press
Taylor, A. E. 1928. *A Commentary on Plato's Timaeus*, Oxford: Oxford University Press
Todd, R. B. 2012. *Themistius on Aristotle Physics 1–3*, London: Bristol Classical Press
Van Melsen, A. G. 1960. *From Atomos to Atom*, New York: Harper & Brothers
Vitelli, H. 1887–8. *Philoponus in Aristotelis Physica Commentaria*, Berlin: Reimer
Vlastos, G. 1950. 'The physical theory of Anaxagoras', *Philosophical Review* 59: 31–57
Wagner, H. 1967. *Aristoteles: Physikvorlesung*, Berlin: Akademie Verlag
Waterlow, S. 1982. *Nature, Change, and Agency in Aristotle's Physics: A Philosophical Study*, Oxford: Clarendon Press
Wicksteed, P. and Cornford, F. M. 1934. *Aristotle: The Physics, Books I–V*, Cambridge, MA: Harvard University Press
Wieland, W. 1970. *Die aristotelische Physik*, 2nd edn, Göttingen: Vandenhoeck & Ruprecht; trans. into Italian as *La fisica di Aristotele*, Bologna: Il Mulino, 1993
1975. 'Aristotle's *Physics* and the problem of inquiry into principles', in *Articles on Aristotle*, vol. I: *Science*, eds. J. Barnes, M. Schofield and R. Sorabji, London: Duckworth: 127–40; also ed., 'Das Problem der Prinzipienforschung und die aristotelische Physik', *Kant-Studien* 52 (1960–1): 206–19
1992. *Die aristotelische Physik*, 3rd edn, Göttingen: Vandenhoeck & Ruprecht.
Williams, C. J. F. 1982. *Aristotle's De Generatione et Corruptione: Translated with Notes*, Oxford: Clarendon Press

Index locorum

De generatione animalium

De generatione et corruptione

De interpretatione

De Melisso, Xenophane et Gorgia

De partibus animalium

Meteorologica